Characterization and Modeling to Control Sintered Ceramic Microstructures and Properties

Technical Resources

Journal of the American Ceramic Society

www.ceramicjournal.org

With the highest impact factor of any ceramics-specific journal, the *Journal of the American Ceramic Society* is the world's leading source of published research in ceramics and related materials sciences.

Contents include ceramic processing science; electric and dielectic properties; mechanical, thermal and chemical properties; microstructure and phase equilibria; and much more.

Journal of the American Ceramic Society is abstracted/indexed in Chemical Abstracts, Ceramic Abstracts, Cambridge Scientific, ISI's Web of Science, Science Citation Index, Chemistry Citation Index, Materials Science Citation Index, Reaction Citation Index, Current Contents/ Physical, Chemical and Earth Sciences, Current Contents/Engineering, Computing and Technology, plus more.

View abstracts of all content from 1997 through the current issue at no charge at www.ceramicjournal.org. Subscribers receive full-text access to online content.

Published monthly in print and online. Annual subscription runs from January through December. ISSN 0002-7820

International Journal of Applied Ceramic Technology

www.ceramics.org/act

Launched in January 2004, *International Journal of Applied Ceramic Technology* is a must read for engineers, scientists,and companies using or exploring the use of engineered ceramics in product and commercial applications.

Led by an editorial board of experts from industry, government and universities, *International Journal of Applied Ceramic Technology* is a peer-reviewed publication that provides the latest information on fuel cells, nanotechnology, ceramic armor, thermal and environmental barrier coatings, functional materials, ceramic matrix composites, biomaterials, and other cutting-edge topics.

Go to www.ceramics.org/act to see the current issue's table of contents listing state-of-the-art coverage of important topics by internationally recognized leaders.

Published quarterly. Annual subscription runs from January through December.
ISSN 1546-542X

American Ceramic Society Bulletin

www.ceramicbulletin.org

The *American Ceramic Society Bulletin*, is a must-read publication devoted to current and emerging developments in materials, manufacturing processes, instrumentation, equipment, and systems impacting the global ceramics and glass industries.

The *Bulletin* is written primarily for key specifiers of products and services: researchers, engineers, other technical personnel and corporate managers involved in the research, development and manufacture of ceramic and glass products. Membership in The American Ceramic Society includes a subscription to the *Bulletin*, including online access.

Published monthly in print and online, the December issue includes the annual *ceramicSOURCE* company directory and buyer's guide. ISSN 0002-7812

Ceramic Engineering and Science Proceedings (CESP)

www.ceramics.org/cesp

Practical and effective solutions for manufacturing and processing issues are offered by industry experts. CESP includes five issues per year: Glass Problems, Whitewares & Materials, Advanced Ceramics and Composites, Porcelain Enamel. Annual subscription runs from January to December. ISSN 0196-6219

ACerS-NIST Phase Equilibria Diagrams CD-ROM Database Version 3.0

www.ceramics.org/phasecd

The ACerS-NIST Phase Equilibria Diagrams CD-ROM Database Version 3.0 contains more than 19,000 diagrams previously published in 20 phase volumes produced as part of the ACerS-NIST Phase Equilibria Diagrams Program: Volumes I through XIII; Annuals 91, 92 and 93; High Tc Superconductors I & II; Zirconium & Zirconia Systems; and Electronic Ceramics I. The CD-ROM includes full commentaries and interactive capabilities.

Characterization and Modeling to Control Sintered Ceramic Microstructures and Properties

Proceedings of the 106th Annual Meeting of The American Ceramic Society, Indianapolis, Indiana, USA (2004)

Editor
C.B. DiAntonio

Published by
The American Ceramic Society
PO Box 6136
Westerville, Ohio 43086-6136
www.ceramics.org

Characterization and Modeling to Control Sintered Ceramic Microstructures and Properties

For information on ordering titles published by The American Ceramic Society, or to request a publications catalog, please call 614-794-5890, or visit our website at www.ceramics.org

ISBN 1-57498-178-1

Contents

Engineering Ceramic Processes and Microstructures

Preface

Predicting and controlling sintered ceramic microstructure and properties is a topic of considerable worldwide interest, and one that is becoming increasingly more important with multi-materials integration (e.g., microelectronics and other composite structures). Characterization, modeling and process control play an important role in the development and manufacture of traditional and advanced ceramics, particularly for process understanding and quality control/assurance. Practical characterization techniques are continually being developed and refined to better understand and control ceramic powder processing, and to support predictive modeling. Additionally, international collaborations are currently in place to develop practical characterization technology and international standards that will impact global ceramic manufacturing well into the 21st century.

This Ceramic Transactions volume is a collection of selected papers that integrates a variety of crucial areas in the development and understanding of sintering and densification. The papers include examinations into the characterization and modeling of sintered ceramic microstructures and the corresponding properties of those microstructures. This provides an increased understanding and expanded knowledge base that can be used by researchers and engineers. The papers were originally presented at the Characterization and Modeling to Control Sintered Ceramic Microstructure and Properties Symposium held during the 106th Annual Meeting of The American Ceramic Society in Indianapolis, Indiana, April 18-21, 2004. All of the papers in this volume were peer-reviewed.

The symposium was designed to provide a forum that integrated research in characterization and modeling to advance the science of ceramic/composite sintering. Densification, shape deformation, and microstructure evolution during sintering was addressed. Contributions on new developments and emerging characterization technologies, practical applications of traditional characterization techniques, and characterization techniques adapted from allied fields are presented. Three major topical areas emphasized in the symposium are presented in this collection: 1) characterization to support modeling (e.g., measuring constitutive properties) and to test and validate model predictions; 2) modeling to predict and control densification, shape deformation, microstructure, and properties; and 3) applications of characterization and modeling to engineer ceramic processes and ceramic microstructures.

I would like to personally thank Kevin G. Ewsuk, Sandia National Laboratories, and Eugene Olevsky, San Diego State University, for their assistance and encouragement in organizing this symposium. I hope that the dissemination of the work presented at the symposium, through this transactions volume, fosters a continued growth in the understanding of the fundamentals as applied to sintering, characterization and modeling of ceramics. I would also like to acknowledge the numerous contributions and support of the speakers, conference session chairs, manuscript reviewers, manuscript authors and The American Ceramic Society officials. Your hard work and devotion to the advancement of science has allowed us to provide this volume to the ceramics community.

C.B. DiAntonio

Characterizing Sintering

Utilizing The Master Sintering Curve to Probe Sintering Mechanisms

D. Lynn Johnson
Department of Materials Science and Engineering
Northwestern University
Phone: 847-491-3584
Fax: 847-491-7820
Email: dl-johnson@northwestern.edu

Abstract

The master sintering curve (MSC) is empirically derived from densification data obtained over a range of heating rates or sintering temperatures. When the proper activation energy is used, all the data converge onto a single curve, the MSC. The activation energy can be estimated readily with just a few dilatometer experiments if it is unknown beforehand. Once established, the MSC makes it possible to predict the final density after arbitrary temperature-time excursions. It is particularly useful when considering alternative sintering methods. The MSC works if a single activation energy controls sintering, but also makes it possible to detect concurrent mechanisms.

Introduction

The ability to predict sintering behavior has been one of the long-term objectives of sintering studies for many decades. The complexity of the sintering process, which depends upon a large array of factors, largely has thwarted attainment of this objective. It is well understood that the sintering behavior of any given material depends upon several characteristics of the powder, including composition, the particle size and size distribution, particle shape, degree of agglomeration, agglomerate size and size distribution, and other factors. One of the principal composition issues is whether there will be a liquid phase present in the sintering temperature range. Compaction method, green density, and uniformity of green density are among the green compact issues that will influence sintering. Sintering temperature obviously has the most effect of any factor, but the heating rate may be an issue also. The atmosphere can have marked effects in some systems. A further complication arises if pressure is applied.

A predictive model would require knowledge of all relevant diffusion coefficients and their temperature dependencies, grain growth kinetics as a function of density, and more. If a liquid phase is present then the solubility of the majority material in the liquid, as well as liquid phase transport information must be known as functions of temperature. Thus, a truly predictive model, wherein sintering can be predicted based upon knowledge of all the relevant factors that might have a significant effect, has yet to be developed. Moreover, for practical purposes, such a model would require too much prior effort to acquire the necessary data. Thus, a more simplified approach is needed, even at the expense of detailed knowledge of every aspect of the process.

From the earliest quantitative sintering studies, over five decades ago, models have been sought that related the sintering rate to the characteristics of the particles, the compact, the atmosphere, and the temperature.[1-5] Simplified geometries were employed to readily identify driving forces, mass transport paths, and geometric factors as sintering proceeded. Attempts were made to extend these simple models into compacts of both spherical and non-spherical powders with limited success.

Many of the simplified models can be derived from the following general model for densification by a combination of grain boundary and volume diffusion[6] by employing the geometric and other assumptions of the simplified models; the master sintering curve also derives from this model:

$$\frac{d\rho}{\rho\,dt} = \frac{3\gamma\Omega_a}{k_B T}\left[\frac{\delta D_b \Gamma_b}{G^4} + \frac{D_v \Gamma_v}{G^3}\right] \tag{1}$$

where ρ = density

γ = specific surface free energy, assumed to be isotropic (which, of course, it is not in general true for solids)

Ω_a = atomic volume

k_B = Boltzmann constant

G = mean grain diameter

δ = thickness of the region of enhanced diffusion at the grain boundary

D_b = grain boundary diffusion coefficient

D_v = volume diffusion coefficient

Γ_b = geometric factor for grain boundary diffusion $= \dfrac{\alpha C_k C_b}{C_\lambda C_a C_h}$ (2)

Γ_v = geometric factor for volume diffusion $= \dfrac{\alpha C_k C_v}{C_\lambda C_a C_h}$ (3)

The geometric factors change continuously as sintering proceeds. They can be understood on the basis of the DeHoff model of a grain in a sintering compact.[7] Each grain is considered to be an irregular polyhedron defined by the grain boundaries between the grain and its nearest neighbors. The polyhedron, in turn, comprises pyramids with a common apex at the center of the grain, the bases of which are defined at the grain boundaries. The polyhedron is extended into the pores so that the total volume of the compact is included in the sum of all the polyhedra.

Each of the factors in Eqs. (2) and (3) relates a particular geometric factor for sintering to the mean grain size, G, as follows:

$\nabla\mu = \dfrac{\alpha K}{\lambda}$ = gradient in chemical potential

$\lambda = C_\lambda G$ = maximum distance of diffusion

$K = -\dfrac{C_k}{G}$ = curvature at the pore or neck surface

$$\delta \frac{L_b}{2} = \delta C_b G \quad \text{= area for grain boundary diffusion}$$

$$A_v = C_v G^2 \quad \text{= area for volume diffusion}$$

$$S^b = C_a G^2 \quad \text{= grain boundary area at the base of the pyramid}$$

$$h = C_h G \quad \text{= height of the pyramid}$$

Experimental determination of Γ_b where the diffusion coefficient was know has shown that this factor changes by two or three orders of magnitude as density increases from 50% to 85% of theoretical, and then begins to increase with further densification. [6]

Su[8] proposed that Eq. (1), simplified by excluding volume diffusion, be rearranged and integrated as follows, recognizing that both G and Γ_b will evolve with density:

$$\frac{k_B}{3\gamma\Omega\delta D_o}\int\frac{[G(\rho)]^4}{\rho\Gamma_b(\rho)}d\rho = \int\frac{1}{T}\exp\left(-\frac{Q_b}{RT}\right)dt \tag{4}$$

This places all constants and difficult to quantify geometry-sensitive terms on the left, leaving only the temperature dependence on the right. He then suggested that the density be determined empirically as a function of the right hand side. Thus density is plotted as a function of Θ, where

$$\Theta = \int\frac{1}{T}\exp\left(-\frac{Q_b}{RT}\right)dt \tag{5}$$

This became known as the master sintering curve. In essence, a sintering process model is derived from the data. Note that no assumption is made about the dependence of temperature on time.

Once the MSC has been determined, sintering can be predicted for arbitrary temperature-time excursions, providing the conditions under which the MSC were determined are not violated (see below).

Under isothermal conditions, Eq. (5) becomes

$$\Theta = \frac{t}{T}\exp\left(-\frac{Q_b}{RT}\right) \tag{6}$$

The time required to reach a particular density under isothermal heating can be computed using Eq. (6) once the MSC has been determined, but of course the non-isothermal portion of the heating cycle must be taken into account using Eq. (5).

It should be noted that Chu, et al., employed an equation similar to Eq. (4) in their study of constant heating rate sintering of ZnO, but focused their attention on the left hand side.[9]

The data provided by Chu, et al. provided a test of the MSC. The raw data are shown in Figure 1 as extracted manually from their Figure 2; the excellent quality of the data is apparent. The MSC, computed with an activation energy of 300 kJ/mol, is displayed in Figure 2. The data appear to fall onto a single curve. (In this and all subsequent plots Θ is in units of s/°K.)

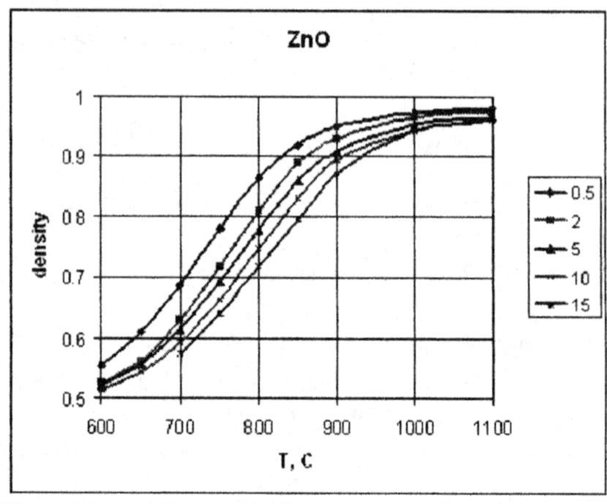

Figure 1. Data of Chu, et. al.[9] for the sintering of ZnO at the constant heating rates shown in the legend.

It turns out that Shercliff and Ashby used a similar approach to develop a process model for age hardening of aluminum alloys, noting that the large number of factors that govern age hardening precludes development of a comprehensive model.[10] Although several sub-models of various aspects of age hardening, such as precipitation, precipitate coarsening, solid solution strengthening, and dislocation interactions with precipitates are well understood., these contain microscopic constants that cannot be predicted with the precision needed. They utilized a detailed calibration procedure to establish the process model. A simpler approach is satisfactory for the MSC.

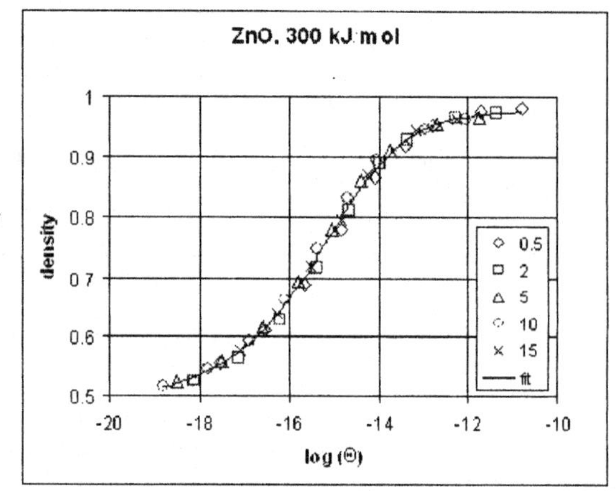

Figure 2. Master sintering curve for the data of Figure 1 assuming an activation energy of 300 kJ/mol.

Shercliff and Ashby call **Θ** (their ***J***) the "kinetic strength," using the nomenclature of an earlier paper on heat-affected zones in welding.[11] They suggest it characterizes the total number of diffusive atom jumps made during the heat treating process. They used it specifically to characterize particle coarsening during heat treating of aluminum alloys. It is quite likely that the concept can be applied to a number of kinetic processes in which a single thermally activated mechanism governs the evolution of the structure and properties of a material.

As useful as the concept is, not all sintering is expected to be described by a master sintering curve since certain conditions must be satisfied. First of all, the green compacts must be consistent in powder characteristics, compaction process, and green density. However, a master sintering surface could be generated for a range of green densities, with green density as the third axis of the plot. Second, a single activation energy must govern sintering under all conditions of interest. This is discussed below.

Finding the master sintering curve

The MSC is generated from measured densities as a function of Θ, using isothermal or non-isothermal methods as long as the temperature as a function of time is known from the beginning to the end, including the early stages of cooling if significant densification could occur during cooling. The easiest method of obtaining data for the MSC is constant heating rate (CHR) sintering with a dilatometer. Three to five runs are made over a range of heating rates, preferably with the highest rate at least a factor of ten greater than the lowest. The final densities are measured, and the density at previous times is computed from the linear shrinkage data. Isothermal sintering over a range of temperatures also can be used as long as shrinkage during the non-isothermal portion of the run is recorded accurately. Finally, individual pellets can be utilized if a suitable dilatometer is unavailable, with the entire heating cycle recorded, but this entails considerably more work, and there generally is more scatter than with dilatometry.

Once a set of data is obtained, the value of Θ is computed for each data point, using a known or assumed value of the activation energy. The data then are plotted as density vs. $\log(\Theta)$, maintaining the identity of each heating rate (for CHR) or temperature (for isothermal) and the dispersion of the data is assessed. If the scatter appears to be random, then the activation energy may be appropriate. If there is a systematic dispersion of curves as a function of heating rate or temperature, then the activation energy should be adjusted. The new values of Θ are computed, the dispersion is assessed, and the process repeated until satisfactory results are obtained.

Of course, the dispersion is best assessed quantitatively, rather than visually. First the data are lumped into a single set, now ignoring the individuality of the runs. A function is then fitted to the data, the predicted density and residual is computed at each data point, and the mean square residual computed. Su and Johnson used polynomial fits to the data, but Teng, et al., showed that the following equation is more powerful.[12] It yields a sigmoidal curve based on the five adjustable parameters, ρ_o, Θ_o, a, b, and c.

$$\rho = \rho_o + \frac{a}{\left[1 + \exp\left(-\frac{\log(\Theta) - \log(\Theta_o)}{b}\right)\right]^c} \quad (7)$$

Here ρ is density, ρ_o is the initial density (really the lower asymptote), a is the difference between the upper and lower asymptotes, Θ_o is the value of Θ at the point of inflection of the curve, and b and c exercise further control over the shape of the curve.

Teng, et al., have developed a Visual Basic macro in Microsoft Excel (MasterCurve) that automates the computations required to find the MSC using this equation, greatly reducing the required effort. The program finds the activation energy value that minimizes the mean square residual of the data.[1] As an

---

[1] The Microsoft Excel macro is available from M.-H. Teng at mhteng@ccms.ntu.edu.tw or the present author at dl-johnson@northwestern.edu.

example, consider the Chu, et al. data referred to above.

MasterCurve computes Θ for each data point and finds the best fit of Eq. (7) to the results for a series of values of the activation energy, Q. Figure 3 (black circles) shows the mean square residual as a function of Q, the minimum of which identifies the best value of Q, 350 kJ/mol. This is somewhat greater than the 310 kJ/mol obtained by Su and Johnson using polynomial fits to the data. It also is greater than the 210 kJ/mol assumed by Chu, et al.,[9] and the 276 ± 13 kJ/mol observed by Gupta and Coble.[13]

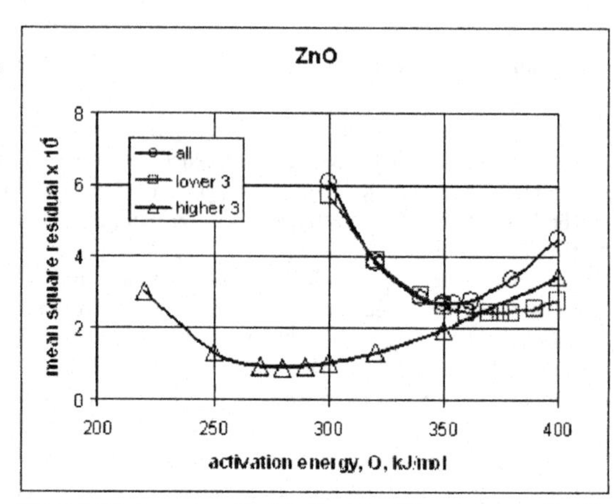

Figure 3. Mean square residual for the data of Chu, et al. using Eq. (7) as the fitting equation. Black circles = all data, red squares = lowest three heating rates, blue triangles = highest three heating rates.

Figure 4 shows the MSC for all the data with Q = 350 kJ/mol. Superficially, the fitted curve appears to adequately describe the data. However, Chu, et al. observed a coarsening process that did not produce densification, and Su and Johnson confirmed this by obtaining the MSC for the three highest heating rates. They proposed that surface diffusion, with an activation energy less than that for densification, is responsible for this coarsening.

The presence of surface diffusion will result in retardation of densification, since the resulting coarsening will lessen the driving force for sintering and increase the mass transport distance. If the activation energy is less than that for the densification mechanism, the retardation effect will diminish with both increased temperature and increased sintering. Thus, there will be more retardation for lower heating rates, which will result in an apparent activation energy greater than that for the densification mechanism.

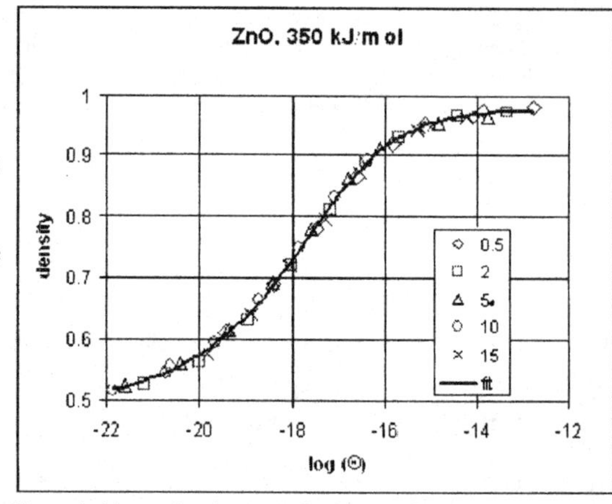

Figure 4. Master sintering curve for the data of Chu et al. with Q = 350 kJ/mol. The legend gives the heating rates in °C/minute.

Using Eq. (7) for the highest three heating rates gives the curve denoted by blue triangles in Figure 3. The activation energy in this case is 280 kJ/mole, about the same as that reported by Gupta and Coble.

Figure 5 shows the MSC for this value of Q. While the fit for the highest heating rates is better than that in Figure 3, there is a clear shift of the two lower heating rate curves to higher values of log (Θ) in the intermediate density ranges. Interestingly, these two curves cross over the MSC at high density. This higher density at lower heating rates was noted by Chu et al. Figure 3 shows the results for the lowest three heating rates. The apparent activation energy is 370 kJ/mol. The MSC is quite similar to that shown in Figure 4.

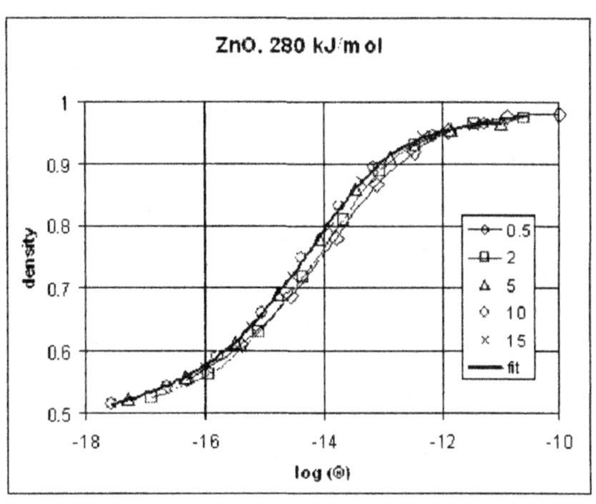

Figure 5. Master sintering curve for the data of Chu et al. with Q = 280 kJ/mol. The legend gives the heating rates in °C/minute.

As a second example, Barringer et al. sintered monodispersed spherical TiO_2 powders with a mean particle size of 0.30 μm under both isothermal and CHR conditions.[14] The data were extracted manually from their Figures 3 and 12.

MasterCurve gave activation energies of 272 and 278 kJ/mol for the CHR and isothermal data, respectively. There was significantly more scatter in the isothermal data, at least partly due to random errors in picking the data from the small figures in the published paper. There is undoubtedly some bias in the isothermal data because of uncertainty in defining the zero of time, as noted by the authors, since considerable shrinkage occurred before the isothermal temperature was reached. This underscores the need to include measurement of densification during the entire heating cycle. Figure 6 shows the CHR data using the average activation energy, 275 kJ/mol, which is virtually indistinguishable from the curve obtained at 272 kJ/mol.

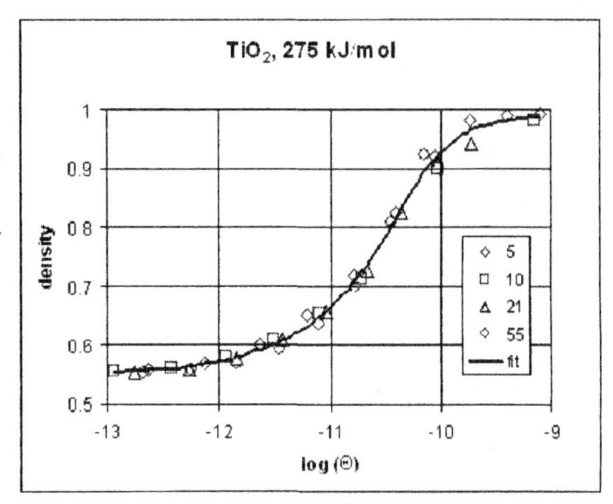

Figure 6. Master sintering curve for CHR sintering of monodispersed spherical TiO_2.[14] The legend gives the heating rates in °C/minute.

An and Johnson showed that a master sintering surface can be generated for uniaxial hot pressing.[15] The third axis is the pressure. Figure 7 shows the results for Sumitomo AKP-30 alumina which had been cold pressed at 300 MPa with 3% polyvinyl butyral binder and pre-sintered at 650° C. The temperature was increased at 10°/min to 1500°, and the density was determined from the final densities and dilatometer records. Constant pressures of 0, 6.9, 13.8, 20.7, 27.6, and 34.5 MPa were maintained during heating. The curves in Figure 7 are constant pressure contours of the fitted surface at these pressures, while the symbols are the measured data points.

Utilizing the master sintering curve

Once the master sintering curve (or the master sintering surface, with green density or pressure as the third axis) has been obtained, the density for arbitrary temperature-time excursions can be predicted. Su and Johnson showed examples of this. As a further example, the CHR data of Barringer et al., Figure 6, can be used to predict their isothermal data.

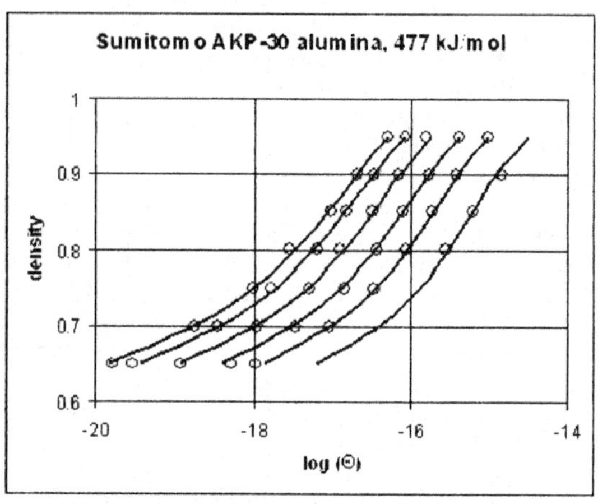

Figure 7. Pressure-assisted master sintering surface for Sumitomo AKP-30 alumina.[15] Solid curves are constant pressure contours from 0 to 34.5 MPa in 6.9 MPa increments, reading from right to left. Symbols represent the measured data.

The predicted densification curves for holding at various temperatures after heating at 55° C/min to the specified temperature are shown as the solid curves in Figure 8. While the fit is not as good as one might hope, the trends are clearly predicted. Deviations from the predicted curves at the higher temperatures are likely caused by the uncertainty in defining the zero of time.

The greatest deviation is for the lowest temperature, which would not be surprising if there is any mass transport by surface diffusion. The heating rates used were not low enough to detect such a contribution from the CHR data.

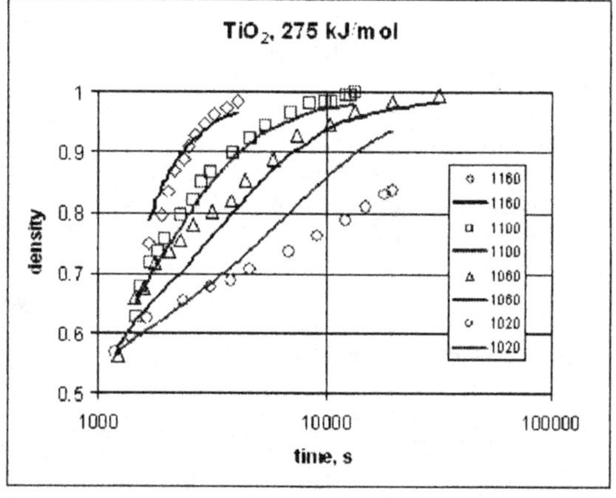

Figure 8. Isothermal densification of monodispersed spherical TiO_2 powder. Solid lines are predictions and symbols are reported values. Isothermal temperatures are given in °C in the legend.

The MSC can be used to predict worst-case warping during firing, or to assist in determining maximum heating rates to avoid cracking if the temperature as a function of position in a piece is known. The prediction is based on the temperature difference between one side and the other of the piece, assuming that each surface follows the MSC. It is a worst case because the geometric constraints imposed by the bulk of the specimen will inhibit shrinkage on the hot side and enhance it on the cooler side, thus causing small deviations from the MSC in the direction to lessen the difference in density, and thence of warping.

The data of Figure 4 can be used as an example. Figure 9 shows the predicted radius of curvature as a function of hot face temperature for a slab of ZnO 10 mm thick for a constant temperature difference of 10° at a heating rate of 50°/min, assuming there is no constraint in the sintering of the faces. Because of constraints, the true curve will show somewhat less curvature, and there probably will be much less increase in the radius of curvature beyond the minimum.

The possibility of fast firing can be explored once the MSC has been established. Suppose a short pass-through furnace has a Gaussian temperature profile with a peak temperature of 1150° and σ of 50 mm, which yields a temperature of 718° at a distance of 50 mm on each side of the position of the peak temperature. How fast can small specimens be passed through this furnace and still achieve density greater than 95% of theoretical? Again using the data of Figure 4, we see that $\Theta=10^{-15}$ is required to achieve 95% of theoretical density. Figure 10 shows the predicted final density as a function of pass-through speed. Of course, thermal conductivity of the specimen becomes an important issue here, but it is neglected in this figure by assuming sufficiently small specimens.

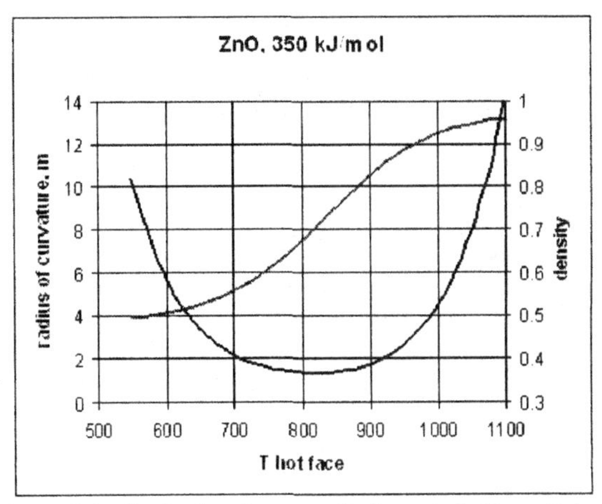

Figure 9. Predicted radius of curvature (black curve) of a 10 mm thick slab of ZnO when heated at 50° C/min with a 10° difference between the hot and cooler face assuming no constraints in the densification of the faces. The red curve is the density of the hot face.

The pressure-assisted master sintering surface (PMSS) can be utilized to predict the density during hot pressing at constant pressure under arbitrary heating cycles. Thus the required kinetic strength (Θ) to obtain a desired density is known for any pressure within the range tested. Alternatively, if the density and Θ are known for a complex pressing situation, the effective pressure can be inferred.

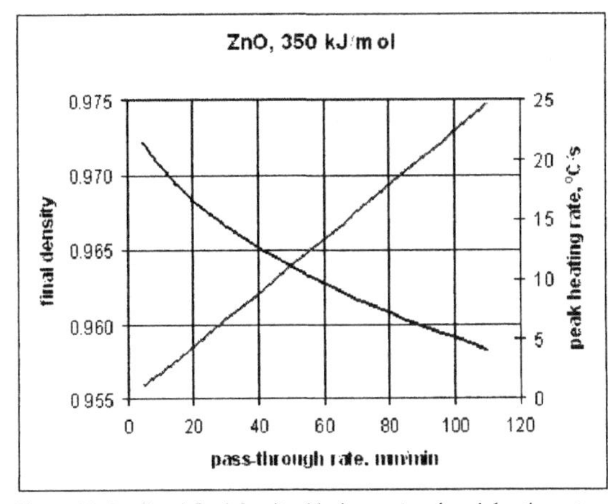

Figure 10. Predicted final density (black curve) and peak heating rate (red curve) for ZnO passed through a short furnace with a peak temperature of 1150° C.

Other uses of the MSC can be envisioned.

Complications

One of the biggest complications is precision and accuracy in the data. It probably is not surprising that the greater the scatter, the more difficult it is to obtain the MSC. A few outlier points near the ends of the curves can have major effects. Statistical analysis of the residuals is advised to identify outliers. Teng's MasterCurve provides residuals for this analysis.

More than a single operative mechanism, and activation energy, invalidates the MSC. Surface diffusion is the most likely second mechanism, as discussed above, and can be handled fairly well if it does not influence the results for the higher heating rates. In some cases, the apparent activation energy needed to get convergence of the data is much too large. This usually is indicative of surface diffusion interference over all the data space. The resulting MSC may not be a good predictor for all heating schedules.

Conclusions

The master sintering curve (MSC) can be obtained from a series of constant heating rate or isothermal sintering runs in which the temperature program and densification both are carefully measured. It can be extended into the third dimension of green density or hot pressing pressure to form a master sintering surface (MSS). Once obtained, the MSC or MSS can be used to predict the density of similar compacts over arbitrary thermal processing conditions, as long as these conditions are consistent with the regime in which the curve (or surface) was obtained. The existence of a secondary mechanism, such as surface diffusion, can be detected.

References

1. G. C. Kuczynski, Trans. Am. Inst. Mining Met. Eng., **185** [2] 169-78 (1949).

2. W. D. Kingery and M. Berg, J. Appl. Phys., **26** [10] 1205-12 (1955).

3. R. L. Coble, J. Amer. Ceram. Soc., **41** [2] 55-62 (1958).

4. R. L. Coble, J. Appl. Phys. **32** [5] 787-92 (1961); **36** 2327 (C) (1965).

5. D. Lynn Johnson, J. Appl. Phys. 40 [1] 192-200 (1969).

6. James D. Hansen, Richard P. Rusin, Mao-Hua Teng, and D. Lynn Johnson, J. Am. Ceram. Soc., **75** [5] 1129-35 (1992).

7. R. T. DeHoff, *Sintering and Heterogeneous Catalysis*, Edited by G. C. Kuczynski, A. E. Miller, and G. A. Sargent, Plenum Press, New York, (1984), pp. 23-34.

8. Hunghai Su and D. Lynn Johnson, J. Amer. Ceram. Soc., **79** [12] 3211-17 (1996).

9. M.-Y. Chu, M. N. Rahaman, L. C. De Jonghe, and R. J. Brook, J. Amer. Ceram. Soc., **74** [6] 1217-25 (1991).

10. H. R. Shercliff and M. F. Ashby, Acta Metall. Mater., **38** [10] 1803-12 (1990).

11. J. C. Ion, K. E. Easterling, and M. F. Ashby, Acta Metall. **32** [11] 1949 (1984).

12. Mao-Hua Teng, Yi-Chun Lai, and Ying-Tien Chen, Western Pacific Earth Sciences, **2** [2] 171-180 (2002).

13. T. K. Gupta and R. L. Coble, J. Amer. Ceram. Soc., **51** [9] 521-25 (1968).

14. E. A. Barringer, R. J. Brook, and H. K. Bowen, *Sintering and Heterogeneous Catalysis*, Edited by G. C. Kuczynski, A. E. Miller, and G. A. Sargent, Plenum Press, New York, (1984), pp. 1-21.

15. K. An and D. Lynn Johnson, J. Mat. Sci., **37** 1-5 (2002).

CONTROLLED AND PREDICTED CERAMIC SINTERING THROUGH MASTER SINTERING CURVE THEORY

Christopher B. DiAntonio
Sandia National Laboratories
Ceramics and Glass Processing Department, 14192
P.O. Box 5800, MS0959
Albuquerque, New Mexico 87185-0100

K.G. Ewsuk
Sandia National Laboratories
Ceramic Materials Department, 1843
Advanced Materials Laboratory, Suite 100
1001 University Blvd SE
Albuquerque, New Mexico 87106

ABSTRACT

Understanding, controlling, and predicting ceramic material sintering behavior is crucial to reproducible manufacturing of high performance specialty ceramic components. The microstructure and properties of a finished ceramic component are highly dependent on the firing process (i.e., sintering and densification). Master sintering curve theory provides a means for the construction of an empirically derived single sintering curve for a ceramic powder system from measured densification data. The measured densification behavior, of a particular ceramic component, is collected over a range of heating rates or sintering temperatures and when combined with the calculated effective densification activation energy for the system all of the data converge onto a single sintering curve. Originally developed for materials that exhibit isotropic sintering behavior and that densify by solid state sintering (Al_2O_3, ZrO_2), MSC theory has recently been extended to construct master sintering curves for an anistropically densifying low temperature co-fire ceramic (LTCC) system and a nanocrystalline ZnO powder. This work will provide some of the fundamental results from the implementation of master sintering curve theory to an anisotropically densifying system and nanocrystalline powder.

Sandia National Laboratories is a multi-program laboratory operated by Sandia Corporation, a Lockheed Martin Company for the United States Department of Energy under contract No. DE-AC04-94AL8500.

INTRODUCTION

One of the many challenges still facing the sintering community is to develop the ability to make quantitative predictions and devise strategies from these predications that allow for control over the sintering process of ceramic powder systems. Ceramic component yields from the firing process are, in most cases, less than optimal, and uncomfortably unpredictable. To develop more robust processes to more reproducibly manufacture reliable products, a far better understanding of the critical control parameters for the densification processes is mandatory. Existing methods cannot predict quantitatively, based on actual measured sintering characteristics, the sintering kinetics of most ceramic powder systems that are of manufacturing interest. Implementation of Master Sintering Curve (MSC) theory provides a systematic approach to control and design ceramic sintering and represents a science-based technology to predict densification during sintering [1].

In order to demonstrate the functionality of the theory and the ability to construct a MSC this work introduces how this tool has been adapted to provide the ability to predict and control the densification behavior for an anisotropically densifying low temperature co-fire ceramic system and a nanocrystalline ZnO powder system.

MASTER SINTERING CURVE (MSC) THEORY

The basic concept of MSC theory is to provide a characteristic measure of sinterability over a given density range which results in a single empirical densification curve that is independent of heating history. This relatively new concept takes advantage of the parameters used in the sintering rate equation by separating terms relating microstructure and temperature to opposite sides of the equation [1,2]. The two sides are then related to each other experimentally. The concept of the MSC is general and this generalization is possible because, often for a given powder and green-body process, the geometric parameters of microstructure are independent of the thermal sintering path. Ashby previously developed similar sintering diagrams; however this technique is considered fundamentally different [3].

The formulation and construction of master sintering curves and subsequent density relationship is rooted in theory developed from the combined stage sintering model.[2] This model is not confined by the segments described by the individual stage models but instead extends the analysis of sintering. The individual stage models contain idealized geometric considerations that do not represent the entire sintering process [4-8]. The instantaneous linear shrinkage in the combined stage model is given as:

$$-\frac{dL}{Ldt}=\frac{d\rho}{3\rho dt}=\frac{\gamma\Omega}{kT}\left(\frac{\delta D_b\Gamma_b}{G^4}+\frac{D_v\Gamma_v}{G^3}\right) \tag{1}$$

where γ = surface energy, G = mean grain diameter, Ω = atomic volume, δ = width of the grain boundary, D_V = volume diffusion coefficient, k = Boltzmann constant, D_b = grain boundary diffusion coefficient, Γ = "Lumped" scaling parameters, dt = change in time, dL = change in length, dρ = change in density, and T = absolute temperature.

The master sintering equations are then developed through a series of assumptions and the subsequent rearrangement of equation 1. The first assumption of the model is that the sintering body undergoes isotropic linear shrinkage. It is important to note that although the original MSC theory is applied to isotropically densifying materials it has been extended for application on anisotropically densifying materials. Therefore this assumption is not necessary and is validated through this work. Second, it is assumed that a single dominant diffusion mechanism exists in the system where volume or grain boundary diffusion dominates the sintering process. The MSC theory may not apply when surface or vapor transport dominates the diffusion process or when exaggerated grain growth occurs, however, it may be able to indicate the presence of these factors. Third, the master sintering curve is a single-valued function of density where G(ρ) (mean grain diameter) and Γ(ρ) (scaling parameters) are a function only of the density of the material and not the time-temperature profile. The developed master sintering curve is therefore unique for a given powder, green microstructure and green density. Any changes to the particle size distribution, average particle size, initial pore-size distribution, and/or particle packing properties will modify the green microstructure and ultimately the master sintering curve. Equation 1, under these assumptions, can be simplified, rearranged and integrated into the following forms:

$$\int_0^t \frac{\gamma\Omega D_0}{kT}\exp\left(-\frac{Q}{RT}\right)dt = \int_{\rho_0}^{\rho} \frac{(G(\rho))^n}{3\rho\Gamma(\rho)}d\rho \qquad (2)$$

$$\Phi_{(\rho)} \equiv \frac{k}{\gamma\Omega D_0}\int_{\rho_0}^{\rho} \frac{(G(\rho))^n}{3\rho\Gamma(\rho)}d\rho \qquad (3)$$

$$\theta(t, T_{(t)}) \equiv \int_0^t \frac{1}{T}\exp\left(-\frac{Q}{RT}\right)dt \qquad (4)$$

thus,

$$\Phi_{(\rho)} \equiv \theta(t, T_{(t)}) \qquad (5)$$

where Q = apparent activation energy, R = gas constant, for volume diffusion $D_O = (D_V)_O$ and n=3, and for grain-boundary diffusion $D_O = (\delta D_b)_O$ and n=4.

Equation 3 includes all microstructural and materials properties, except for Q (activation energy). Equation 4 depends only on the activation energy and the time-temperature profile. The relationship between ρ and Φ(ρ) is defined as the master sintering curve, where density is a function of θ(t,T) (the time-temperature trajectory), and takes advantage of the equality in equation 5 [1]. In its simplest implementation and form the master sintering curve is generated through dilatometric linear shrinkage measurements using a range of constant heating rates. The density as a function of time is calculated from the dilatometer data and plotted versus θ(t,T), using a specific effective activation energy for densification. A generic example of this is shown schematically in Figure 1.

The effective activation energy of densification (Q_{eff}) for this reaction rate phenomena was calculated using Arrhenius reaction theory and Arrhenius plots [9,10]. Determination of the activation energy is complicated by the fact that the densification rate changes as a function of temperature and grain size. It must be assumed that the relationship between density and grain size remains constant over the temperature range in order to make an unambiguous estimate of the activation energy. Separating and rearranging the sintering rate equation into temperature-dependent, grain-size dependent, and density dependent quantities [10] accomplishes this:

$$\ln\left(T\frac{dT}{dt}\frac{d\rho}{dT}\right) = -\frac{Q}{RT} + \ln[f(\rho)] + \ln A - n\ln d \qquad (6)$$

where T=absolute temperature, t=time, ρ=density, Q=activation energy, R=gas constant, f(ρ)=variation of density, A=material parameter, n=grain size power law exponent, and d=grain size.

The activation energy is then determined through an Arrhenius plot of the left hand side of equation 6 versus 1/T, provided that the data points are taken at a constant value of density. The use of multiple heating rate experiments produces the necessary values of constant density. The effective activation energy (Q_{eff}) is calculated from the slope of each fit linear line [Q =(R)(Calculated Slope)]. If the correct value of Q_{eff} has been calculated and the mechanism for densification has not changed, all of the data will converge to a single curve, the master sintering curve.

From the measured and combined data the master sintering curve can be formulated and constructed. It is obvious from Figure 1 that each of the heating rate densification curves have collapsed onto one master sintering path (indicated by the average value) where the instantaneous density is a function of the log Θ [log(sec/K)] parameter. The curves are in good agreement over the entire density range studied. This is a good indicator that one diffusion or densification mechanism is dominant throughout the entire sintering treatment.

MASTER SINTERING CURVE- CASE STUDIES

Low Temperature Co-Fire Ceramic (LTCC) System

Low temperature co-fire ceramic (LTCC) electronic packaging technology is presently being used to produce components for telemetry, satellites/radar, MEM systems, etc. The LTCC system provides a technology option for designers of interconnect packaging that offers parallel processing, excellent dielectric isolation, and high layer count circuitry, as well as, high performance conductors, and simple, inexpensive processing. It combines the benefits of both high temperature co-fire ceramics and thick film technologies by providing for a high density, high reliability, high performance, and low cost interconnect package. Functional components have been produced but not without some process reproducibility and control issues. Root causes and solutions to co-firing design, materials, and process control problems will be critical to the future success and application of LTCC packaging technology.

The focus of this work is to examine the anisotropic sintering/densification of a Dupont 951 low-temperature cofire glass/ceramic dielectric tape, one of the components typically incorporated into LTCC electronic packaging technology. This low temperature 951 Green Tape™ is a gold and silver compatible, low coefficient of thermal expansion, high strength glass/ceramic tape. A specific issue in the manufacturing of an LTCC device is the ability to control and predict the sintering behavior of this glass/ceramic component. This ability is critical to reproducible processing of reliable parts. For this reason a sintering model has been developed that uses master sintering curve (MSC) theory to predict density/shrinkage as a function of sintering time and temperature. Implementation of MSC theory provides for a more systematic approach to design and control sintering. In tape cast ceramic substrates, shrinkage is often found to be anisotropic. For example, in high solids loading aqueous alumina suspensions, the shrinkages in the transverse and casting direction vary by more than 12% (typically 13.5% in the casting direction and 15.6% in the transverse direction) [11,12]. Traditional tape casting fabrication of LTCC tape results in anisotropic shrinkage during sintering. Thus allowing, for the first time, application of MSC theory to a system that exhibits anisotropic densification. Dilatometric examination of the shrinkage along two anisotropic axes and application of the master sintering theory produced a master sintering curve for the densification of this material under arbitrary temperature-time excursions. This provides a means to predict final densities as a function of sintering time and temperature. It also allows assessment of lot-to-lot (materials) and run-to-run (process) variability in LTCC manufacturing.

Sintering studies have been conducted on Dupont 951 low temperature cofire glass/ceramic tape through dilatometry of the linear shrinkage during densification. A Perkin-Elmer Thermomechanical Analyzer (TMA), series 7, and Pyris Analysis software for Windows, version 3.81, ©1996 – Perkin-Elmer Instruments LLC, was used to measure the shrinkage of the tape samples as a function of temperature and time. The linear shrinkage was measured using multiple heating rate experiments of 5, 10, 15, 20, 25, and 30°C/min, with a maximum sintering temperature of 950°C. The samples were air quenched at 950°C by removal from the furnace. Due to the anisotropic nature of this shrinkage, and lack of a bi-directional TMA, Z-direction (thickness) and X-Y direction (in-plane) measurements were taken on separate samples and results averaged.

From the measured and combined data the master sintering curve can be formulated and constructed. Application of the master sintering curve theory, described previously, produces the master sintering curve for the LTCC tape, as shown in Figure 2. It is obvious from this plot that each of the heating rate densification curves have collapsed onto one master sintering path (indicated by the average value) where the instantaneous density is a function of the log Θ [log(sec/K)] parameter. The curves are in good agreement over the entire density range studied. The effective activation energy for the dominant densification mechanism in the Dupont 951 LTCC glass/ceramic tape has been determined through Arrhenius reaction rate theory to be 317 ± 38 kJ/mol.

Master sintering curve theory has been successful applied to the anisotropic densification of an LTCC glass/ceramic tape (≈3% shrinkage difference between Z-direction and X,Y-directions). It is now possible to predict and control the densification behavior of the material under arbitrary temperature-time excursions.

Nanocrystalline Zinc Oxide Powder System

Nanomaterials have potential for improving science and engineering as a whole, particularly; improved catalysis, dispersion, transparency, surface smoothness, and gloss, as well as finer abrasiveness, and ceramic toughness in applications ranging from high-tech coatings to fuel cells and multi-layer ceramic capacitors (MLCCs) [13]. Critical to the future success and application of nano-powder material systems and the technology that incorporates them is to address the root causes and solutions to sintering, co-sintering and process control problems. A specific issue in the manufacturing of any ceramic device or application is the ability to control and predict the sintering behavior. For this reason, this work examines master sintering curve (MSC) theory to determine its applicability to be able to predict and aid in controlling the sintering behavior of nano-sized ceramic powders. It was anticipated that MSC theory, if applicable to nanomaterials sintering, could provide a more systematic approach to design and control this sintering.

The focus then is to first verify the MSC theory for a nanocrystalline ZnO* (>99.00% purity) powder system. The physical properties of the powder include; an average particle size of 48nm, a specific surface area of $30m^2/gr$ (BET), a zincite (hexagonal) crystal phase and an elongated morphology. The powder is commercially formed using a Physical Vapor Synthesis (PVS) process.

---

* Nanophase Technologies Corp., 1310 Marquette Dr., Romeoville, IL 60446.

Dilatometer linear shrinkage measurements of isostatically pressed cylindrical powder compacts were taken in the length direction using a Netzsch-Differential-Dilatometer 402ED** (subsequent shrinkage measurements verified isotropic shrinkage in the diameter for all systems examined). The measurements were taken using multiple constant heating rates of 5, 10, 20, and 30°C/minute after the organics, which were used during forming, were pyrolyzed from the samples.

From the measured and combined data the master sintering curve can be formulated and constructed. The average activation energy, determined by an Arrhenius plot, from Arrhenius reaction rate theory, is 268±25kJ/mol for nano-ZnO, over the density range of 65 to 90% of theoretical, respectively. Application of the master sintering curve theory, described previously, produces the master sintering curve for this system, as shown in Figure 3. Each of the heating rate densification curves have collapsed onto one master sintering path (indicated by the average value) where the instantaneous density is a function of the log Θ [log(sec/K)] parameter. The curves are in good agreement over the entire density range studied, 65 to 90% of theoretical density, where the highest fraction of linear shrinkage and densification occurs. This is a good indicator that one diffusion or densification mechanism is dominant through this range during the overall sintering treatment.

CONCLUSION

The ability of the MSC to predict final density based on the time-temperature profile during sintering has been proven by dilatometric testing of multiple samples at multiple heating rates and subsequent sintered density analysis. Sintering is typically accomplished through the interaction of several mechanisms. The identification of the specific mechanisms that are active is complicated. For this reason the effective activation energy for the dominant densification mechanism for each powder system has been determined through Arrhenius reaction rate theory.

It is now possible to predict and control the densification behavior of both an anistropically densifying and nanocrystalline system. Thus providing another method to model differential densification and materials interactions in order to minimize shape deformation and microstructure heterogeneity (e.g. cracks and porosity). These flaws can degrade process reproducibility, product performance and reliability. This has been accomplished without the difficult task of quantifying the microstructural evolution during sintering and its subsequent effects on sintering kinetics.

Another interesting point to note is that sample sets can be sintered using temperature-time profiles that include dwell times and multiple ramp rates. The MSC however was not constructed using these types of time-temperature profiles, thus, further providing an indication of the overall flexibility of the design and theory.

** Netzsch Instruments, Inc..

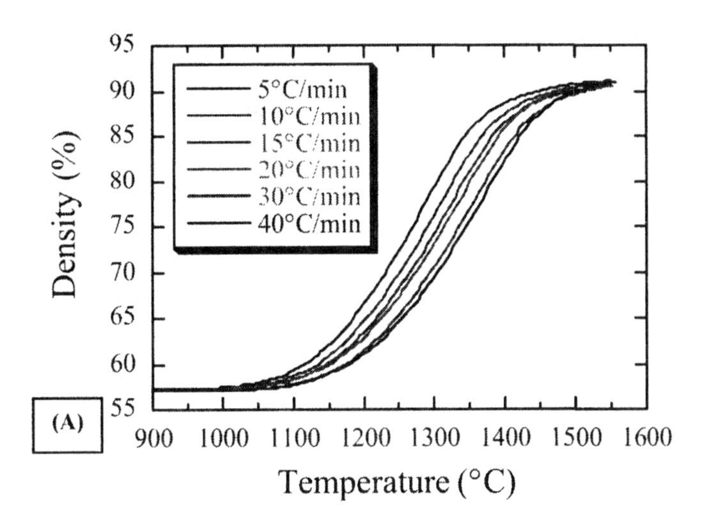

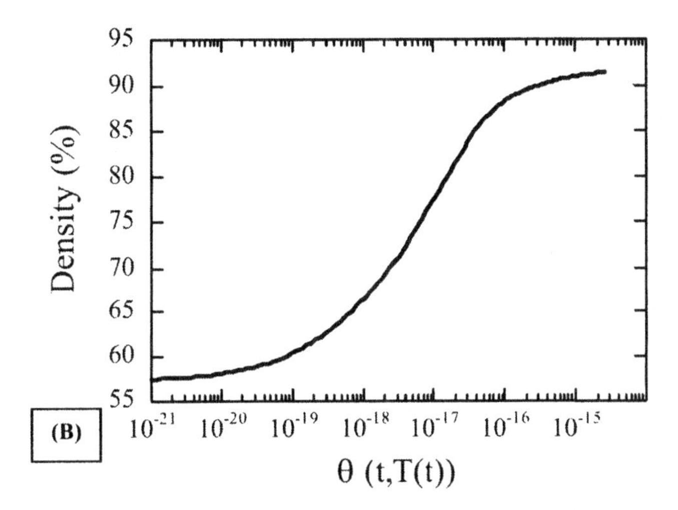

Figure 1. (A) Density as a function of sintering temperature from dilatometric data, (B) Calculated master sintering curve, based on measured linear shrinkage data and subsequent density curve, for a generic ceramic system, respectively.

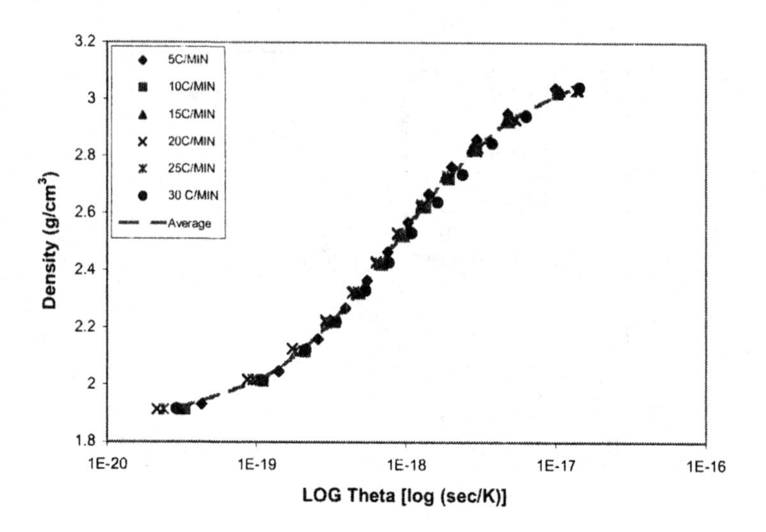

Figure 2. Master sintering curve (MSC) for Dupont 951 LTCC glass/ceramic tape using an effective activation energy for densification of 317±38 kJ/mol.

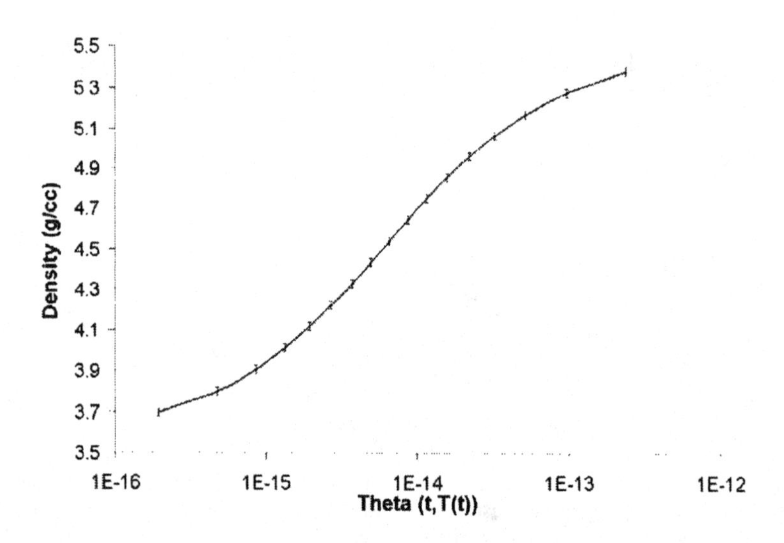

Figure 3. Average master sintering curve for nano-ZnO using an effective activation energy of densification of 268±25kJ/mol.

REFERENCES

[1]H. Su and D.L. Johnson, "Master Sintering Curve: A Practical Approach to Sintering," *J. Am. Ceram. Soc.*, 79 [12] 3211-17 (1996).

[2]J. Hansen, R.P. Rusin, M. Teng, and D.L. Johnson, "Combined-Stage Sintering Model," *J. Am. Ceram. Soc.*, 75 [5] 1129-35 (1992).

[3]M.F. Ashby, "A First Report on Sintering Diagrams," *Acta Metall.*, 22 [3] 275-89 (1974).

[4]G.C. Kuczynski, "Self-Diffusion in Sintering of Metallic Particles," *Trans. Am. Inst. Mining Met. Eng.*, 185 [2] 169-78 (1949).

[5]W.D. Kingery and M. Berg, "Study of Initial Stages of Sintering Solids by Viscous Flow, Evaporation-Condensation, and Self-Diffusion," *J. Appl. Phys.*, 26 [10] 1205-12 (1955).

[6]R.L. Coble, "Sintering of Crystalline Solids I: Intermediate and Final Stage Diffusion Models," *J. Appl. Phys.*, 32 [5] 787-92 (1961).

[7]J. Zhao and M.P. Harmer, "Sintering Kinetics for a Model Final-Stage Microstructure: A Study in Al_2O_3," *Philos. Mag. Lett.*, 63 [1] 7-14 (1991).

[8]D.L. Johnson, "New Method of Obtaining Volume, Grain-Boundary, and Surface Diffusion Coefficients from Sintering Data," *J. Appl. Phys.*, 40 [1] 192-200 (1969).

[9]W.S. Young and I.B. Cutler, "Initial Sintering with Constant Rates of Heating," *J. Am. Ceram. Soc.*, 53 [12] 659-63 (1970).

[10]J. Wang and R. Raj, "Estimate of the Activation Energies for Boundary Diffusion from Rate-Controlled Sintering of Pure Alumina, and Alumina Doped with Zirconia or Titania," *J. Am. Ceram. Soc.*, 73 [5] 1172-75 (1990).

[11]B. Schwartz, "Review of Multilayer Ceramics for Microelectronic Packaging," *J. Phys. Chem. Solids*, 45 [10] 1051-54 (1984).

[12]P. Markondeya Raj and W. Cannon, "Anisotropic Shrinkage in Tape-Cast Alumina: Role of Processing Parameters and Particle Shape," *J. Am. Ceram. Soc.*, 82 [10] 2619-25 (1999).

[13]C. C. Koch, "Nanostructured Materials: Processing, Properties and Potential Applications," Noyes Publications, William Andrew Publishing, Norwich, New York, ©2002.

SINTERING DAMAGE DURING MULTI-MATERIAL SINTERING

Terry J. Garino
Sandia National Laboratories
P.O. Box 5800, MS-1411
Albuquerque, NM 87185-1411

ABSTRACT

To generate data for comparison with the predictions of continuum sintering models for multi-material systems, several types of concentric cylinder samples were sintered to produce damage during sintering. The samples consisted of an outer ring of pressed ceramic powder (alumina or zinc oxide), the center of which was either fully or partially filled with a cylinder that consisted of either the same powder pressed to a higher green density (fully filled) or of previously densified 99% alumina (fully or partially filled). In addition, slots of various lengths were cut in some of the rings, from the outer surface parallel to the cylinder axis, which were then fully filled with dense alumina center cylinders and sintered. The types of sintering damage produced as the shrinkage of the rings was constrained by the center cylinders which shrank less or not at all, included shape deformation, cracking and possible density gradient formation. Comparisons of shrinkage measurements on rings fully filled with dense alumina center cylinders indicated that while the presence of the center cylinder increased the thickness and width shrinkage for both materials, the overall densification of the rings was impeded due to the decrease in circumferential shrinkage. This effect was more severe for the zinc oxide rings. The shape of the cross sections of the rings that were sintered either fully or partially filled with dense alumina center cylinders also showed differences depending on their composition.

INTRODUCTION

The co-firing of multi-material structures has been investigated and performed on a commercial scale for a number of years. Examples of co-fired multi-material systems include low-temperature co-fired ceramic (LTCC) packages and solid-oxide fuel cells (SOFCs). These devices are fabricated from powders of the various materials and then co-sintered to achieve the desired degree of densification for each of the materials. However, incompatibility in the shrinkage behavior of the various materials in such a system, caused by their different free sintering shrinkage rates, generates stresses that can in turn affect the co-sintering behavior. In addition to altering of the shrinkage rates of the materials, these stresses can produce sintering damage such as shape deformation, formation of density gradients and the formation of cracks. A continuum, finite-element model that would predict the behavior during sintering of multi-material structures is now under development at Sandia National Laboratories.[1]

The purpose of this study was to prepare and sinter samples with several geometrically simple shapes that exhibit sintering damage so that, in the future, these results can be compared to the predictions of the continuum model. All samples were based on the concentric cylinder geometry that was used in previous work.[2] Incompatibility in sintering shrinkage is produced in the samples by having the green density in the outer ring be lower than that of the center cylinder so that the shrinkage of the ring is constrained by the inner cylinder. The different green densities were achieved by varying the pressure used to press the powder or, to maximize the incompatibility, by using a previously densified cylinder of 99% alumina.

Several modifications of the simple concentric cylinder arrangement were also investigated as shown in Figure 1. Figure 1a shows the basic concentric cylinder with 1.27 cm ID (0.5") and 2.54 cm OD (1.0") with both cylinders having the same thickness, nominally 0.63 cm (0.25"). The second type of sample, shown in Fig. 1b, is the same as the first except a slot has been cut part way through the outer ring at one location. In this case, the effect of slot length was studied. The third type, Fig. 1c, is a non-symmetrical arrangement where the top half of the ring is filled with a center cylinder. Figure 1d shows the final type of sample where the center third of the ring is filled with a center cylinder. Center cylinders of higher green density were used only for the fully filled case, 1a, whereas dense alumina center cylinders were used for all four types.

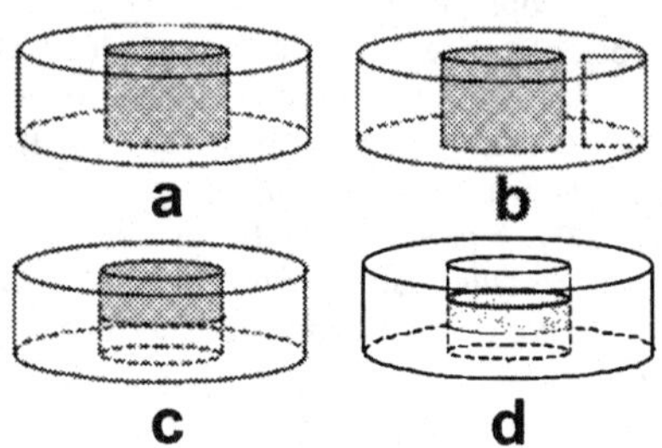

Figure 1. The four types of concentric cylinder samples that were prepared and sintered: a) completely filled ring, b) completely filled ring with slot, c) top half filled ring and d) center-third filled ring.

PROCEDURE

The samples were made from powders of ZnO (0.2 μm, Aldrich Chemical Co., St. Louis, MO) or Al_2O_3 (A16SG, 0.4 μm, Alcoa World Chemicals, Leetsdale, PA). The powders were dispersed in deionized water with a polyelectrolyte dispersant (Darvan C, R.T. Vanderbilt Co., Norwalk, CT) and an ultrasonic horn. Aqueous PVA (15,000 MW, Fluka, Ronkonkoma, NY) and PEG (Compound 20M, Union Carbide Chemicals and Plastics Co., Danbury, CT) binder solutions were added to give 1.5 wt% of each relative to the weight of the powder. After drying, the powders were granulated and passed through a 150 μm sieve.

Cylindrical rings (1.27 cm ID (0.5"), 2.54 cm OD (1.0") and 0.63 cm (0.25") in thickness) of ZnO and Al_2O_3 were uniaxially pressed. The pressure was adjusted to so that samples with two different green densities were produced. For zinc oxide, 31 MPa (4.5 ksi) produced 47% relative green density (GD) and 138 MPa (20 ksi) produced 57%. For alumina, 34 MPa (5 ksi) produced 51% relative GD and 138 MPa (20 ksi) produced 57%. Center cylinders (1.27 cm OD (0.5") and 0.63 cm (0.25") in thickness) of zinc oxide were also pressed to 57% relative GD.

After pyrolysis of the binder, ZnO rings were filled with either the ZnO center cylinders or with pieces of sintered 99% alumina rods (Coors) that either completely filled the rings, or filled only the bottom half or the center third of the ring. The same arrangements of the dense 99% alumina rod sections were used with the alumina rings. Also, for both zinc oxide and alumina rings, a diamond saw was used to cut a slot parallel to the cylinder axis that started at the outer surface of the ring. The length of the slots varied from 1 to 3 mm and these samples were then filled completely with dense alumina rod pieces. The ZnO samples were heated at 5°C/min to temperatures from 750°C to 1000°C, with no hold, before reheating to the next higher temperature. The alumina samples were sintered by heating them at 10°C/min to 1400°C,

1450°C and then 1500°C with no hold. The samples were then reheated to 1500°C and held for 2 hours. An outer ring and an inner cylinder of each GD were also sintered separately for both materials, i.e. without any inner cylinder inserted in a ring. The linear shrinkage of ZnO samples with the two different green densities were measured during heating at 5°C/min using an *in situ* video system.[2] The shrinkage in diameter and thickness of the cylindrical samples was measured using calipers and optical micrographs were taken of the cross sections of some of the samples.

RESULTS AND DISCUSSION

The free sintering linear shrinkage during heating at 5°C/min of the zinc oxide powder at the two levels of green density used is shown in Fig. 2. Only above ~775°C where the shrinkage is ~10% do the curves separate, with the shrinkage of the lower green density sample being greater throughout the rest of the sintering. Therefore, when the zinc oxide concentric cylinder sample consisting of a high green density (57%) center cylinder surrounded by a low green density (47%) ring was sintered, the center does not constrain the shrinkage of the ring until temperatures above 775°C. The shrinkage data for this sample, shown in Fig. 3, verifies this since at 750°C, the shrinkage in OD, ID, thickness and width (one half of the difference between the OD and the ID) are not significantly different from each other, indicating that the ring has shrunk isotropically to that point. Above that temperature, the various shrinkages diverge, with the ID shrinking the least due to the constraint of the center cylinder. When a dense 99% alumina center cylinder was placed inside a zinc oxide ring, the effect the constraint on the sintering of the ring was much greater, as expected. As shown in Fig. 4, the OD, width and thickness shrinkages are different even at 750°C and they continue to diverge such that at 1000°C, the thickness had shrank 17% whereas the OD only 10%. This constraint was severe enough to decrease the densification of the ring, as shown in Fig. 5, compared to both the free sintering and to when the center cylinder consisted of higher density ZnO. No cracks were observed after sintering any of these samples.

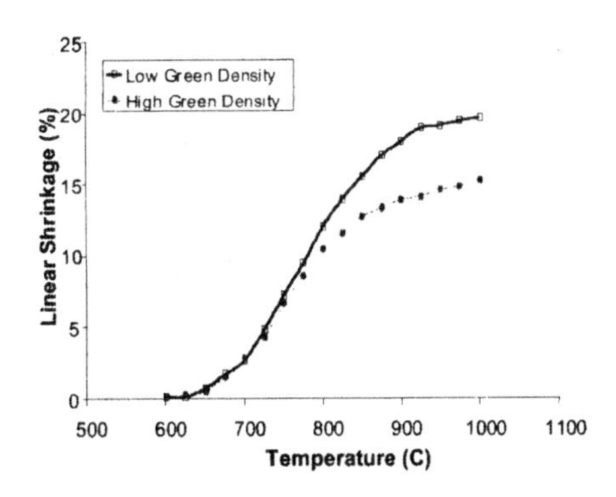

Figure 2. The free shrinkage of the ZnO powder pressed to two different green density values, 47 and 57%.

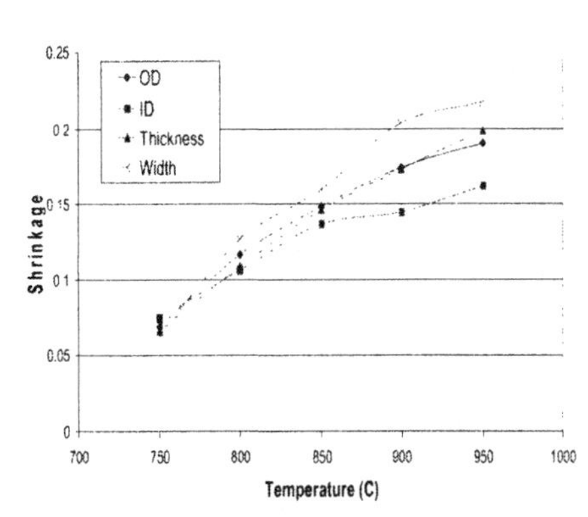

Figure 3. The shrinkage of a ZnO ring with 47% GD that was completely filled with a 57% GD ZnO center cylinder.

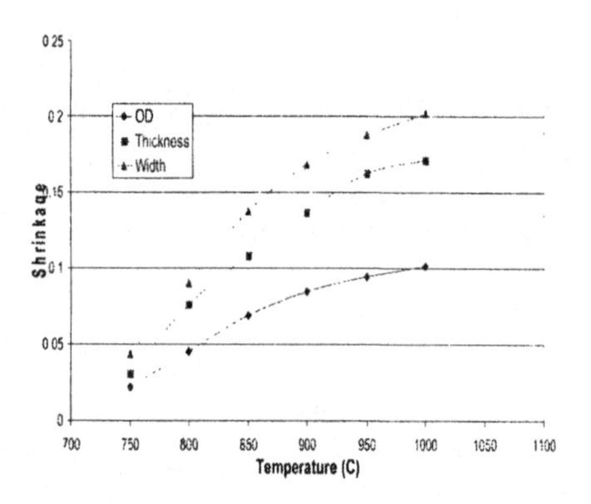

Figure 4. The shrinkage of a ZnO ring with 47% GD that was completely filled with a dense alumina center cylinder.

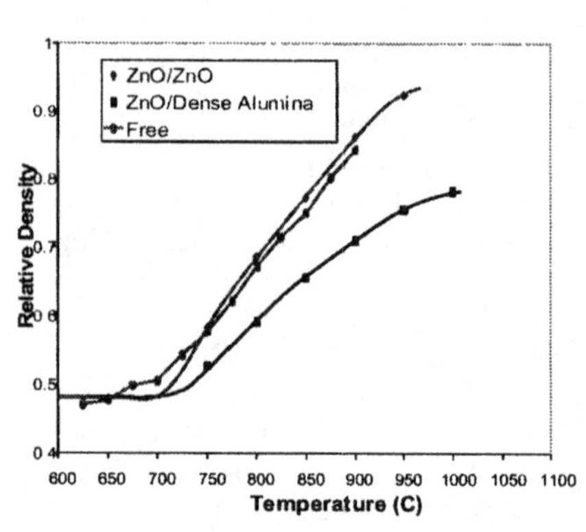

Figure 5. The relative density of ZnO rings with 47% GD, completely filled with 57% GD ZnO or with dense alumina compared to the free sintering of the same material.

The shrinkage data for an alumina ring with 51% green density that was completely filled with a dense cylinder of 99% alumina is shown in Fig. 6. The results were similar to the case of the ZnO ring in that the final shrinkage in width was about twice that of the OD and was several % more than in the thickness direction. However, since the magnitude of shrinkage values for the alumina ring were greater than those for the ZnO ring, the overall densification of the alumina was only slightly less than that of a similar ring that was sintered with its center empty, see Fig. 7.

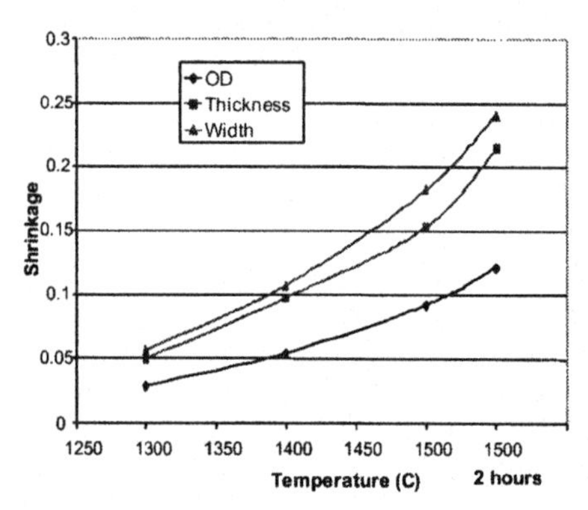

Figure 6. The shrinkage of an alumina ring with 51% GD that was completely filled with a dense alumina center cylinder.

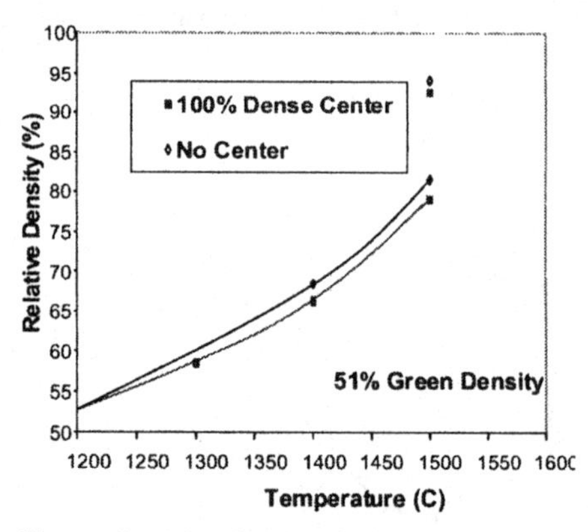

Figure 7. The density during sintering of alumina rings with 51% GD of that were either filled with dense alumina centers or were empty.

The shape of the cross section of alumina (57% GD) and a ZnO (47% GD) rings that were sintered to high density while they were completely filled with dense alumina center cylinders are compared in Fig. 8. The shapes were similar in that the rings were thickest at the inner surface (on the left in both figures) where they contacted the dense center cylinder. For both, the thickness decreased away from the inner surface so that it was constant after about a third of the way towards the outer edge. The magnitude of the maximum increase in thickness was slightly more for the ZnO, 10%, than for the alumina, 8%, but this could be due to the fact the alumina had a higher green density. This effect is due to the frictional force at the interface between the ring and the dense center. The ring applies a compressive force to the center cylinder that produces the friction as the ring tries to slide along the surface as it shrinks in the thickness direction.

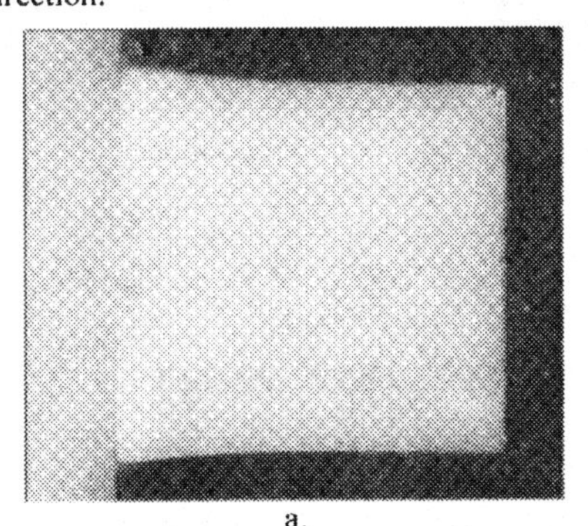

a.

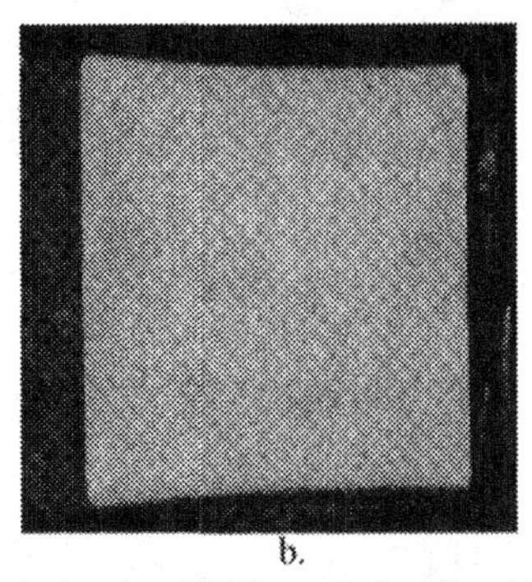

b.

Figure 8. Optical micrographs of the cross sections of a) alumina (57% green density, 1500°C for 2 hr) and b) zinc oxide (47% green density, 1000°C) rings that were sintered completely filled with dense alumina centers.

A series of ZnO rings that were sintered to 950°C completely filled with dense alumina center cylinders after slots of 3.2, 2.2 or 1.2 mm were sawed in the outer edge are shown in Fig. 9a. In all cases, a crack formed at the tip of the slot although the severity of the crack was strongly dependent of the initial slot length. For the sample with the longest slot, the crack propagated all the way across the sample to the inner surface early in the sintering process so that a large gap opened up due to subsequent shrinkage. For the 2.2 mm slot length, the crack grew more than half the way from the tip of the slot to the inner surface. The amount that the crack opened varied approximately linearly from the tip of the slot to the crack tip. Also, as shown in Fig. 9b, a gap opened up between the inner surface of the ring and the center cylinder in the region where slot was located. The sample with the 1.2 mm slot had only a short, narrow crack that is difficult to see in the figure.

Figure 10 shows an optical micrograph of the cross section of an alumina ring sintered (1500°C for 2 hr) with its center half filled with a dense alumina disk. The ring warped severely due to the rotation of the ring cross section towards the inner surface at the top where the ring was empty. The cross sections of rings of ZnO and alumina sintered to high density with their centers half full are compared in Fig. 11 (rotated so that the center cylinder is now in the top half and the outer surface is vertical. Neither ring showed evidence of cracking. This figure shows that for both materials, the top and outer surfaces remained flat whereas as the bottom, where the center was unfilled, is slightly curved upwards near the inner surface, more so for the alumina ring. Also for both materials, the inner surface at the top has pulled away from the surface of the center cylinder and the angle that the outer surface makes with the top is slightly greater than 90° whereas the angle it makes with the bottom is slightly less than 90°. Finally, the bottom portion of the inner surface, which was empty during sintering, was straight except for being curved in the near to the edge of the center cylinder.

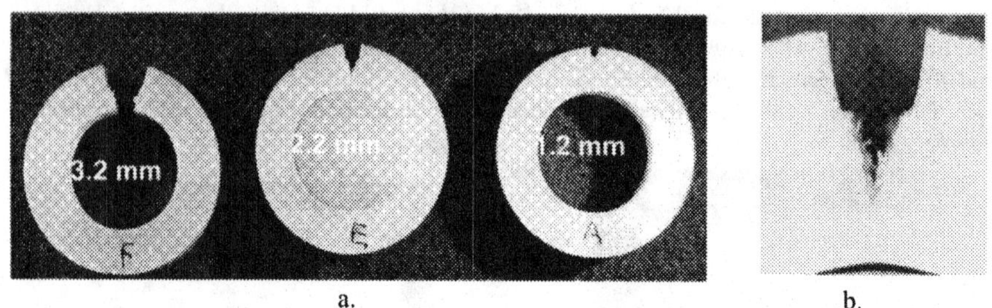

a. b.

Figure 9. Optical micrographs of a) slotted zinc oxide rings (47% green density) with three different initial slot lengths, 1.2, 2.2 and 3.2 mm, that were sintered (1000°C) with dense alumina centers and b) a higher magnification image of the 2.2 mm length slot sample.

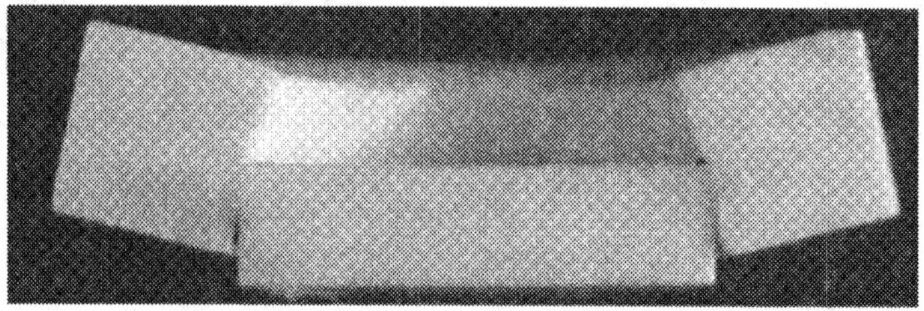

Figure 10. An optical micrograph of the cross section of an alumina ring sintered (1500°C for 2 hr) with its center half filled with a dense alumina disk.

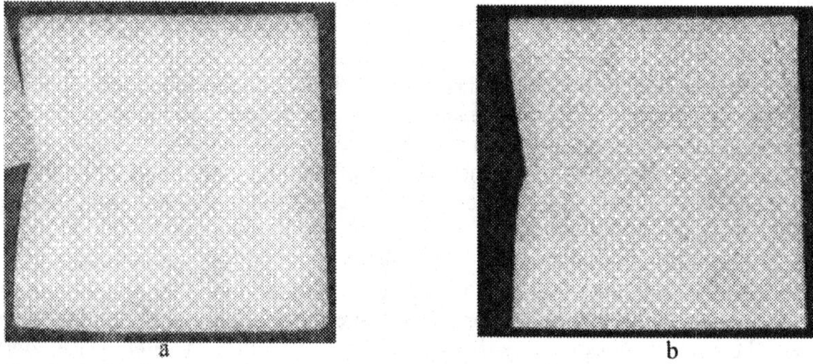

a b

Figure 11. Optical micrographs of the cross sections of a) alumina (57% green density, 1500°C for 2 hr) and b) zinc oxide (47% green density, 1000°C) rings that were sintered half filled with dense alumina centers.

Fig. 12 shows an optical micrograph of the cross section of an alumina ring that was sintered to high density while its center third section was filled with a disk of dense alumina. The top, bottom and outer surfaces of the ring are all flat and the angles between the outer surface and top and bottom are both slightly greater than 90°. The inner surface is curved near the edges of the center cylinder. Even though the inner surfaces near the top and bottom were free to move inwards during sintering, the absence of the constraint in these region did not transfer to the outer surface since, as stated, it remained flat. Therefore, the ring is thinner in the center and thus more shrinkage in the radial direction actually occurred in the center portion of the sample where the constraint of the center cylinder was present than near the top or bottom where it was not. This suggests that gradients in density are developing during the sintering process. Also, several short cracks formed in the ring at the inner surface near the top or the bottom that only penetrated a short distance towards the center.

Figure 12. Optical micrograph of an alumina ring (57% green density, 1500°C for 2 hr) that was filled in the center third with dense alumina.

CONCLUSIONS

Concentric cylinder samples that exhibited various types of sintering damage were produced from zinc oxide and alumina. Somewhat surprisingly, rings with relative green densities as low as 47% could be sintered around dense, non-shrinking, center cylinders without producing cracks unless a slot was pre-cut into the outer surface of the ring. When a slot was present, there was a strong dependence of the extent of cracking on the initial slot length. When the rings were completely filled with the sense center cylinders, the shrinkage in the circumferential, radial and thickness directions differed from each other significantly, with the radial (width) shrinkage being the greatest and the circumferential shrinkage increasing from zero at the inner diameter surface to about half that of the width at the outer surface. Frictional effects at the interface between the dense center cylinder and the inner surface of the ring were seen to cause an increase in the ring thickness in the vicinity of the inner surface. One difference between the behavior of the two materials was that the densification of rings fully filled with dense center cylinders was decreased more for zinc oxide rings than for alumina. There were also subtle differences in the shape of the ring cross sections after sintering between the two materials, especially for the case when they were half filled with the dense center cylinders. The half-filled samples also showed very significant shape deformation or warping as the cross sections of the ring rotated towards the empty portion. Density gradients appear to have been produced in the samples that were sintered with their center third section filled with dense alumina disks since the inner surfaces shrank inwards near the top and bottom but the outer surface remained flat. These results can be compared to predictions of continuum constrained sintering models to see if in fact the models, with the appropriate input parameters for the materials, can accurately predict the types of sintering damage that were produced.

ACKOWLEDGEMENTS

This work was done at Sandia National Laboratories, a multiprogram laboratory operated by Sandia Corporation, a Lockheed Martin Company, for the United States Department of Energy under Contract DE-AC04-94AL85000. The author wishes to thank Dale Zschiesche for technical support.

REFERENCES

1. J. G. Argüello, V. Tikare, T. J. Garino, and M. V. Braginsky, "Three-Dimensional Simulation of Sintering Using a Continuum Modeling Approach," **Sintering, 2003**, Penn State University, September, 2003.
2. T.J. Garino, D.J. Zschiesche and M.L. Howard, "Experimental Studies To Support Multi-Material Sintering Modeling," **Sintering, 2003**, Penn State University, September, 2003.

Modeling Sintering

SINTERING OF ORIENTED PORE-GRAIN STRUCTURES

E.A. Olevsky, B. Kushnarev, and
A.L. Maximenko
San Diego State University
5500 Campanile Dr., San Diego
CA 92182-1323

V. Tikare
Sandia National Laboratories
Materials Theory & Computation
Albuquerque, NM 87185-1411

ABSTRACT

Microstructural evolution during sintering of 2D compacts of elongated particles incorporating ellipsoidal oriented pores at grain junctions is simulated using both a kinetic, Monte Carlo algorithm and a micro-mechanical continuum model. The Monte Carlo model simulates curvature driven grain growth, pore migration, vacancy formation and annihilation. The micro-mechanical continuum model takes into consideration the grain-boundary and surface diffusion mechanisms controlling the evolution of the pore-grain structure. As a result, the kinetics of porosity, grain and pore morphology as well as the shrinkage anisotropy factor are predicted.

INTRODUCTION

Shrinkage anisotropy is one of the most important sintering-related macroscopic phenomena. It's practical evidence is overwhelming and has been reported in numerous publications (see, for example, [1,2,3]). Understanding anisotropy and the development of rigorous models with predictive capabilities could eliminate a major source of distortion problems, which often negate the net-shape character of the process.

The shrinkage anisotropy as a deviation from a self-similar shape evolution is apparently predominantly related to the existence of some structure orientation in the sintered material. It is this idea that has been employed in an overall very limited number of models describing sintering anisotropy. Olevsky *et al.* [4] considered viscous sintering of an anisotropic-porous material thereby reducing particulate structure organization to the embedded texture of ellipsoidal oriented pores. Raj *et al.* [5] and Zavaliangos and Bouvard [6] considered diffusional sintering of ellipsoidal oriented particles at early stages of sintering when interparticle contact area is small. These models are limited by the description of only certain stages of sintering (early [5,6] or late [4]) due to the highly idealized pore-grain structures employed by them. A generalization and refinement of these approaches is possible if a multi-scale simulation with a realistic powder material structure is used.

The present work has a dual purpose: (i) verification of the stochastic mesoscale simulations by a parallel physically-based modeling and (ii) the description of the shrinkage anisotropy for intermediate and late stages (due to the developed grain-boundary area in contrast to the previously developed models [5,6]) of diffusional sintering.

MICRO-MECHANICAL APPROACH

Let us consider a 2-D representative unit cell containing oriented rectangular grains with semi-axes a and c and elliptical also oriented pores located at the grain quadra-junctions (Fig.1). The maximum and minimum curvature radii r_a and r_c of the elliptical pore contour are defined as:

$$r_a = \frac{c_p^2}{a_p};\ r_c = \frac{a_p^2}{c_p} \tag{1}$$

where a_p and c_p are the pores' semi-axes.

The flux of matter caused by the grain boundary diffusion is determined by the chemical potential gradient along the grain boundaries due to the aforementioned normal stresses [7]:

$$J_x^{gb} = -\frac{\delta_{gb} D_{gb}}{kT\Omega}\frac{\partial \mu}{\partial x};\ J_y^{gb} = -\frac{\delta_{gb} D_{gb}}{kT\Omega}\frac{\partial \mu}{\partial y} \tag{2}$$

where D_{gb} is the coefficient of the grain boundary diffusion, δ_{gb} is the grain boundary thickness, k is the Boltzman's constant, T is the absolute temperature, Ω is the atomic volume, and μ is the chemical potential.

Based on (2), the strain rates can be obtained as:

$$\dot{\varepsilon}_y^{f.s.} = -\frac{3}{2}\frac{\delta_{gb} D_{gb}\Omega}{kT}\alpha\frac{1}{(c+c_p)a^2}\left[\frac{1}{r_a}+\frac{1}{a}\sin\frac{\phi}{2}\right];\ \dot{\varepsilon}_x^{f.s.} = -\frac{3}{2}\frac{\delta_{gb} D_{gb}\Omega}{kT}\alpha\frac{1}{(a+a_p)c^2}\left[\frac{1}{r_c}+\frac{1}{c}\sin\frac{\phi}{2}\right] \tag{3}$$

where α is the surface tension, ϕ is the dihedral angle, a and c are the grain semi-axes.

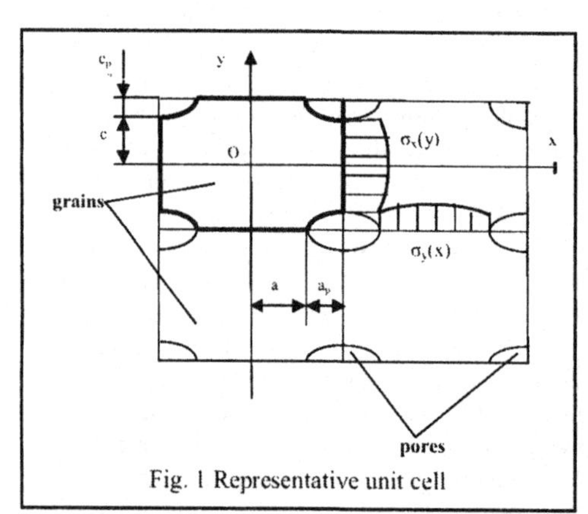

Fig. 1 Representative unit cell

The derived relationships describe the x- and y- shrinkage rates of the representative unite cell, provided that the only mass transport mechanism within grains is the grain boundary diffusion. Both the derived formulas can be utilized if the evolution kinetics of the structure parameters a, c, a_p, and c_p are known.

Thus, the structure parameters under investigation include the grain dimensions (a and c) and the pore semi-axes (a_p and c_p). The evolution of the pore dimensions is related to two mechanisms of mass transport: (i) grain-boundary diffusion, which causes pore collapse by "injecting" additional atoms inside the pore, and (ii) surface diffusion, which redistributes mass along the pore surface without causing any change of the pore area.

A simplified model is used further, which assumes an independent collapse of pores in both directions. This is described by the following kinetic relationships

$$\begin{cases} \dot{a}_{p_c} = -\dfrac{2J_x^{gb}\Omega}{\pi c_p} = -\dfrac{6}{\pi}\dfrac{\delta_{gb} D_{gb}\Omega\alpha}{kT}\dfrac{1}{a\cdot c_p}\left[\dfrac{1}{r_a}+\dfrac{1}{a}\sin\dfrac{\phi}{2}\right] \\ \dot{c}_{p_c} = -\dfrac{2J_y^{gb}\Omega}{\pi a_p} = -\dfrac{6}{\pi}\dfrac{\delta_{gb} D_{gb}\Omega\alpha}{kT}\dfrac{1}{c\cdot a_p}\left[\dfrac{1}{r_c}+\dfrac{1}{c}\sin\dfrac{\phi}{2}\right] \end{cases} \tag{4}$$

Here the subscript "c" indicates that the kinetic parameter is related to the pore collapse due to the grain-boundary diffusion.

The latter equations describe the evolution of the pore dimensions only due to the grain-boundary diffusion. Additionally, one should take into consideration the redistribution of mass due to the surface diffusion.

Following Mullins [8]:

$$J(s) = -\frac{\delta_s D_s}{kT\Omega}\frac{\partial \mu(s)}{\partial s} \tag{5}$$

where δ_s surface diffusive width ($\delta_s = \Omega$ times the number of atoms per unit area), D_s is the coefficient of surface diffusion, s is the arc length along the pore surface, and $\mu(s)$ is the chemical potential defined as:

$$\mu = \frac{\alpha\Omega}{r} \tag{6}$$

where r is the curvature radius.
Hence:

$$\dot{a}_{p_s} = 3\frac{\delta_s D_s \alpha\Omega}{kT}\frac{\left(c_p^2 - a_p^2\right)}{c_p^3 a_p^2}; \quad \dot{c}_{p_s} = -3\frac{\delta_s D_s \alpha\Omega}{kT}\frac{\left(c_p^2 - a_p^2\right)}{c_p^2 a_p^3} \tag{7}$$

The combined kinetics of the pore dimensions can be obtained by the superposition of expressions (4) and (7):

$$\begin{cases} \dot{a}_p = \dot{a}_{p_c} + \dot{a}_{p_s} = -3\frac{\delta_s D_s \Omega\alpha}{kT}\left[\frac{2\Delta c_p^2 a_p^2\left[\frac{1}{r_a} + \frac{1}{a}\sin\frac{\phi}{2}\right] - a\pi\left(c_p^2 - a_p^2\right)}{\pi a_p^2 \cdot c_p^3 \cdot a}\right] \\ \dot{c}_p = \dot{c}_{p_c} + \dot{c}_{p_s} = -3\frac{\delta_s D_s \Omega\alpha}{kT}\left[\frac{2\Delta c_p^2 a_p^2\left[\frac{1}{r_c} + \frac{1}{c}\sin\frac{\phi}{2}\right] + c\pi\left(c_p^2 - a_p^2\right)}{\pi a_p^3 \cdot c_p^2 \cdot c}\right] \end{cases} \tag{8}$$

where $\Delta = \frac{\delta_{gb} D_{gb}}{\delta_s D_s}$ is the ratio characterizing the relative rate of the grain-boundary and the surface diffusion.

The kinetics of the grain size evolution can be determined by subtracting the kinetics of the pore dimensions from the kinetics of the dimensions of the representative unit cell. Following the normalization approach [4], which enables a generalized analysis of the sintering kinetics in the framework of a "master sintering curve" concept, one can normalize the above-mentioned seven equations. In such a case the seven unknown kinetic parameters become functions of the specific dimensionless time τ_s of sintering

$$\tau_s = \left(\frac{\delta_s D_s \alpha\Omega}{kT r_{p_0}^4}\right)t; \quad r_{p_0} = \frac{r_{a_0} + r_{c_0}}{2} = \frac{a_{p_0}^3 + c_{p_0}^3}{2a_{p_0} c_{p_0}} \tag{9}$$

Here r_{p_0} is the initial average dimension of a pore.

MESOSCALE SIMULATIONS

Microstructural evolution during sintering of 2D compacts of elongated particles packed in the arrangement corresponding to Fig. 1 was simulated using a kinetic, Monte Carlo (Potts') model. The initial shape of pores was assumed to be circular. The model presented here is limited to consideration of the following geometry and processes:

- Grain growth by short range diffusion of atoms from one side of the grain boundary to the other;
- Long range diffusion of material to pores by grain boundary diffusion and along pore surfaces by surface diffusion;
- Vacancy annihilation at grain boundaries.

In the model, an ensemble of grain sites and pore sites is allowed to populate a square (in 2-D simulations) or cubic (in 3-D simulations) lattice. The grain sites can assume one of Q distinct, degenerate states, where the individual state is designated by the symbol q and the total number of states in the system is Q, $q_{grain} = [1, 2, \ldots Q]$. The pore sites can assume only one state, $q_{pore} = -1$. Contiguous grain sites of the same state q form a grain and contiguous pore sites form a pore. Grain boundaries exist between neighboring grain sites of different states, q, and pore-grain interfaces exist between neighboring pore and grain sites. The equation of state for these simulations is the sum of all the neighbor interaction energies in the system given by

$$E = \frac{1}{2}\sum_{i=1}^{N}\sum_{j=1}^{8}\left(1-\delta\left(q_i,q_j\right)\right) \tag{10}$$

where N is the total number of sites, δ is the Kronecker delta with $\delta(q_i = q_j) = 1$ and $\delta(q_i \neq q_j) = 0$, q_i is the state of the grain or pore at site i and q_j is the state of the nearest neighbor at site j. As pore sites can assume only one state, $q_{pore} = -1$, there are no pore boundaries and all pores sites coalesce. In contrast, grain sites can assume many different states making grain boundaries possible. Grain growth is simulated using the method developed in previous works [9,10]. First a grain site is chosen at random from the simulation space. Then a new state q is chosen at random from the Q possible states in the system. The grain site is temporarily assigned the new state and the change in energy is evaluated using Eq. (10). Next the standard Metropolis algorithm [11] is used to perform the grain growth step based on Boltzmann statistics. A random number, R, between 0 and 1 is generated. The transition probability, P, is calculated using

$$P = \begin{cases} \exp\left(\dfrac{-\Delta E}{k_B T}\right) & for \quad \Delta E > 0 \\ 1 & for \quad \Delta E \leq 0 \end{cases} \tag{11}$$

where k_B is the Boltzmann constant and T is temperature. If $R \leq P$, then the grain growth step is accepted, if not, the original state is restored. The simulation temperature used for grain growth was $k_B T = 0$, which has been shown to simulate grain growth well [10].

Annihilation is simulated as follows. A straight line is drawn from the isolated pore site through the center of mass of the adjacent grain to the outside boundary of that particle. Next, the isolated pore site and the outside grain site are exchanged with the grain site assuming the q state of the adjacent grain. As the grain boundary length increases, time between annihilations also increases as:

$$t_{anni} = t^{i}_{anni}\left(\frac{L_{gb}}{L^{i}_{gb}}\right)^2 \qquad (12)$$

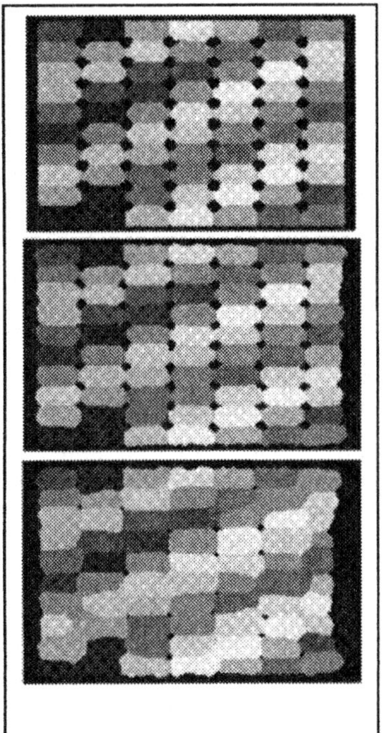

where t_{anni} is the time between annihilation attempts, t^{i}_{anni} is the time between annihilation attempts for the initial grain boundary length L^{i}_{gb} (a_0 or c_0) at the beginning of sintering, and L_{gb} (a or c) is the current grain boundary length. Adjusting the annihilation frequency in this manner simulates uniform annihilation of vacancies along the grain boundary.

Sample simulation results are shown in Fig. 3 in the form of consecutive microstructure images. The starting configuration is perfectly elongated particles with circular pores at all the grain junctions. As the simulation progresses, vacancies are formed at the pore surfaces and diffuse along grain boundaries. They are annihilated at the grain boundaries. This leads to densification of the system. Grain growth does not occur while the pores are present at the junctions and none is expected, as the grain boundaries are not curved. However, once the pores shrink away and four grains meet to form a quadra-junction, they quickly grow in a manner so as to eliminate the quadra-junction and form a triple junction. The simulations show greater shrinkage in the direction of elongation (x-direction) than perpendicular to elongation (in the y-direction).

The sintering strains for the same powder compact obtained from the two models are compared in Fig 5. Despite the assumptions made by the micro-mechanical model, it compares very well with the Monte Carlo model.

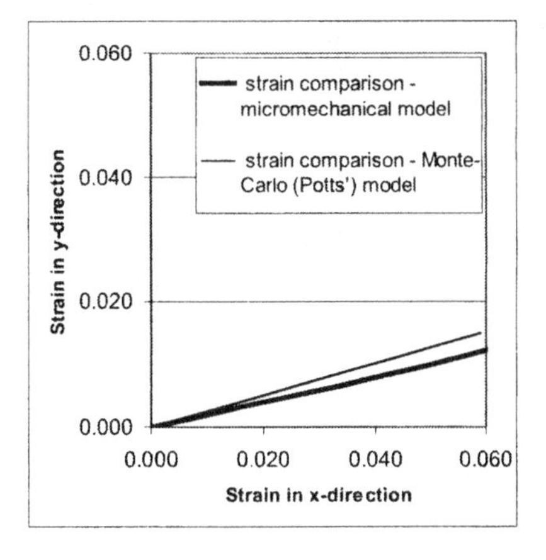

CONCLUSIONS

- A micromechanical model of diffusional sintering of an oriented pore-grain structure has been developed. The model enabled the determination of constitutive and kinetic parameters of the sintered anisotropic material.
- A meso-scale Monte-Carlo model has been used for describing sintering of oriented pore-grain structures. The results of both micromechanical and mesoscale modeling indicate satisfactory agreement.

ACKNOWLEDGMENTS

This work was partially performed at Sandia National Laboratories, a multiprogram laboratory operated by Sandia Corporation, a Lockheed Martin Company, for the USDOE under the Contract DE-AC04-94AL-85000. The support of the NSF Divisions of Manufacturing and Industrial Innovations, Civil and Mechanical Systems, and Materials Research (Grants DMI-9985427, DMI-0354857, CMS-0301115, DMR-0313346) is gratefully acknowledged.

REFERENCES

[1]. Hausner H.H., The linear shrinkage behavior of metal powder compacts during sintering, *Progr. Powder Metall.*, 19, 67-85 (1963)

[2]. Exner H.E., Principles of single phase sintering, *Rev. Powd. Met. Phys. Ceram.*, 1, 7-251 (1979)

[3]. German R.M., Sintering Theory and Practice, John Wiley & Sons, Inc. (1996)

[4]. Olevsky E., Skorohod V., Deformation aspects of anisotropic-porous bodies sintering, *Journal de Physique IV*, C7, 3, 739 (1993)

[5]. Raj P. M., Odulena A., Cannon W.R., Anisotropic shrinkage during sintering of particle-oriented systems—numerical simulation and experimental studies, *Acta Mater.*, 50, 2559–2570 (2002)

[6]. Zvaliangos A., Bouvard D., Mumerical simulation of anisotropy in sintering due to prior compaction, *Int. J. Powder Metall.*, 36, 58-65 (2000)

[7]. Johnson D.L., New method of obtaining volume, grain-boundary, and surface diffusion coefficients from sintering data, *J. Appl. Phys.*, 40, 192-200 (1969)

[8]. Mullins W.W., *J. Appl. Phys.* 28, 333 (1957)

[9]. Anderson M.P., Srolovitz D.J., Grest G.S., and Sahni P.S., Computer simulation of grain growth - I. Kinetics, *Acta Metall.* 32, 783-791 (1984)

[10]. Holm E.A., James A. Glazier, D.J. Srolovitz, G.S. Grest, Effects of lattice anisotropy and temperature on domain growth in the two-dimensional potts model, *Phys. Rev. A*, 43, 2662-2668 (1991)

[11]. Metropolis N., Rosenbluth A.W., Rosenbluth M.N., Teller A.N. and Teller E., Equation of state calculations by fast computing machines, *J. Chem. Phys.*, 21 1087-1092 (1953)

FINITE ELEMENT SIMULATION OF DENSIFICATION AND SHAPE DEFORMATION DURING SINTERING

H. Camacho and J. Castro
Universidad Autónoma de Ciudad Juárez (UACJ), Instituto de Ingeniería y Tecnología (IIT) Chih, México 32310

L. Fuentes and A. García
Centro de Investigación en Materiales Avanzados (CIMAV), Chih, México, 31109

M. E. Fuentes
Universidad Autónoma de Chihuahua, Chih, Ciudad Universitaria México

ABSTRACT

Considering sintering as a viscous process, a stress equilibrium problem is stated in the ANSYS software based on the Finite Element Analysis. In order to take into account the dependence of the constitutive properties on density and temperature, the Scherer Cell Model is implemented into the ANSYS code. The initial finite element model is developed in the initial green body geometry. The thermal program is divided into several steps and, for each one, deformation rates and densification are calculated. For each element, the next step density value is calculated from the densification rate. In order to move from one step to the next one the node displacements are calculated. This way, deformations during sintering modify the shape of the finite element model. As these deformations are the result of the sintering viscous problem, they are a consequence of the stress distribution across the sintering body. This stress distribution is a function of density and temperature gradients as they affect the constitutive properties. In the sintering viscous problem, it is possible to import the load stress resulting from other processes. In the present work, we consider the mismatch of the thermal dilations due to temperature distribution by means of the shear thermal stresses. The thermal hydrostatic stress is not supposed to alter the body shape. A significant expected result is that for lower reference viscosity values, densification and deformation reach higher values. Comparing the thermal stress profile evolution with the shape evolution of the Finite Element Model, it cannot be stated that the sintering body shape totally follows the thermal evolution profile. We believe that this is due to the viscous stress gradient resulting from the temperature and density dependence of sintering constitutive properties.

INTRODUCTION

Sintering is a process that is gaining more and more space in industry. One of the most notable advantages is that this process allows manufacturing many materials without reaching the melting point. Material powder is formed and fired. During firing sintering takes place, grains get closer to each other and the strength of the material is considerably increased. Sintering permits to obtain solid bodies starting from powders. Even though, our civilization has been firing ceramic compact powders for thousands of years, the research activity in this area is still of interest. The ceramic firing process is not totally controlled in industry yet and the optimization for cost decreasing is a constant demand.

In the present work, we focus our attention on solving a macroscopic problem to describe densification and deformation during sintering. An acceptable solution to the macroscopic problem is an application of the theory of continuous media,[1] which requires the use of constitutive equations. Bordia and Scherer[1] have shown that many materials can be considered viscous during sintering. In principle we state a stress equilibrium problem to calculate the deformation rate tensor to determine densification and deformation. A thermal problem is considered and divided into several steps. The thermal stress effect is included in the viscous stress problem. Similar problems have been solved in the literature by other authors: Cantavella,[2] Riedel *et al*,[3-5] and Cedergren.[6] We believe our proposal to be novel in the way that the effect of temperature variation is considered. All problems are solved using the commercial software ANSYS based on Finite Element Analysis.

PROBLEM STATEMENT

The continuum equilibrium equation has the form

$$\frac{\partial \sigma_{ij}}{\partial x_j} = 0 \tag{1}$$

where σ_{ij} represents the stress tensor and x_j the coordinates.

The strain tensor is defined as follow:

$$\varepsilon_{ij} = \frac{1}{2}\left(\frac{\partial u_i}{\partial x_j} + \frac{\partial u_j}{\partial x_i}\right) \tag{2}$$

where u_i is the displacement along the coordinate x_i.

For the present problem, our focus is to predict the deformation rates to describe densification. The deformation rate tensor, $\dot{\varepsilon}_{ij}$, is given in terms of velocity, v_i, as

$$\dot{\varepsilon}_{ij} = \frac{1}{2}\left(\frac{\partial v_i}{\partial x_j} + \frac{\partial v_j}{\partial x_i}\right) = \frac{d}{dt}\varepsilon_{ij} \tag{3}$$

The starting point of the present model is a geometrical figure. Then, this figure is meshed and as a result of doing so a set of nodes appears in the model. Firstly, a transient thermal problem is solved. The temperature distribution is imported in a thermo-elastic problem, $\sigma_{ij} = E_{ijkl}\left(\varepsilon_{kl} + \delta_{ij}\delta_{kl}\alpha_T \Delta T\right)$, where E_{ijkl} is the Elastic modulus tensor, α_T is the thermal coefficient, ΔT is the difference between the actual and the reference temperature and δ_{ij} is the Kronecker delta. We consider that the hydrostatic thermal stresses are not supposed to change the shape of the sintering body (model). Then, in the viscous sintering problem, the shear thermal stresses are imported. This last problem allows us to calculate the deformation rates. Figure 1 shows the flow chart of this procedure.

The viscous constitutive relation can be written as

$$\sigma_{ij} = C_{ijkl}\,\dot{\varepsilon}_{kl} + \delta_{ij}\sigma_s \tag{4}$$

where C_{ijkl} represents the viscous modulus tensor and σ_s the sintering stress. Formally, σ_s is the stress that prevents a sintering body from shrinking. It represents an externally applied hydrostatic stress that would have the same effect on the densification rate as the grain boundaries and pores.

The boundary conditions have the form

$$\sigma_{ij}\big|_{BS} = 0 \qquad (5)$$

and

$$\left|v_i\big|_{SP}\right| = \delta \qquad (6)$$

where *BS* is the boundary surface of the macroscopic cylinder, *SP* represented some selected key points or lines where some displacement or velocity δ is permitted. Equations (1) and (4)-(6) constitute a complete mathematical problem that can have a unique solution if the boundary conditions, Eqs. (5) and (6), are properly stated.

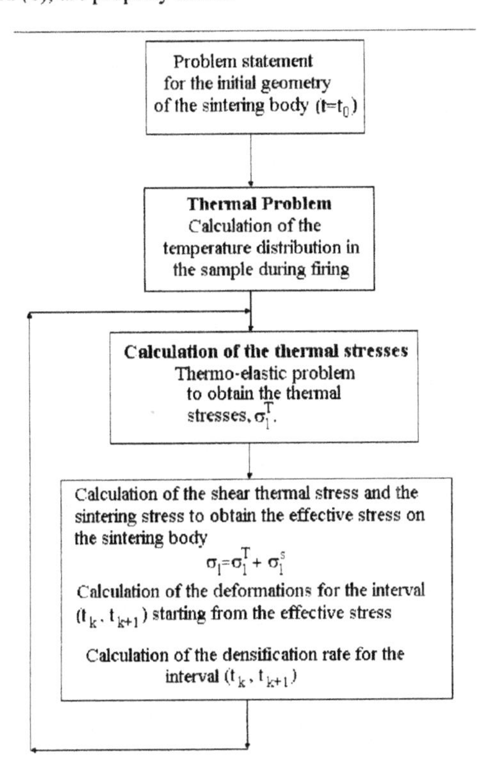

Fig. 1. Flow chart for the calculating procedure.

The calculations are implemented in ANSYS 5.5 Multiphysics. The thermal problem is solved with *SOLID70* element and the thermo-elastic problem is solved with *SOLID45* element (Fig. 1). We based the solution of the viscous problem on the *SOLID45* element as well, because of the viscous and elastic mathematical similarity. For our problem, the constitutive coefficients, C_{ijkl}, E_{ijkl} are functions of temperature and density.[7] In this way, the stress equilibrium of Eq. (1) is calculated. According to the ANSYS methodology, we have the option to supply the corresponding boundary conditions of Eqs. (5) and (6). The ANSYS software has libraries for the different constitutive behaviors; *SOLID45* considers that $\sigma_{ij} = E_{ijkl}\varepsilon_{kl}$. According to the viscous analogy,[1] $E_{ijkl} \leftrightarrow C_{ijkl}$ and $\varepsilon_{kl} \leftrightarrow \dot{\varepsilon}_{kl}$. The term $\delta_{ij}\sigma_s$ in Eq. (4) is not taken into account. Therefore, a final adjustment is required for the problem implementation in ANSYS.

The ANSYS stress is defined as

$$\sigma_{ij}^{A} = C_{ijkl}\,\dot{\varepsilon}_{kl} \tag{7}$$

Using Eqs. (1) and (4)-(6) to express the actual problem, Eq. (5) then is rewritten as

$$\sigma_{ij}\Big|_{BS} = \left(C_{ijkl}\,\dot{\varepsilon}_{kl} + \delta_{ij}\sigma_s\right)\Big|_{BS} = 0 \tag{8}$$

Equation (8) can be rewritten as

$$C_{ijkl}\,\dot{\varepsilon}_{kl}\Big|_{BS} = -\delta_{ij}\sigma_s \tag{9}$$

Sintering stress is a function of temperature and density. Hence, assuming no important temperature and density gradients in each element of the Finite Element mesh, we can assume that

$$\frac{\partial \sigma_{ij}}{\partial x_j} = \frac{\partial}{\partial x_j}\left(C_{ijkl}\,\dot{\varepsilon}_{kl} + \sigma_s\right) = \frac{\partial C_{ijkl}\,\dot{\varepsilon}_{kl}}{\partial x_j} = \frac{\partial \sigma_{ij}^{A}}{\partial x_j} \tag{10}$$

Finally, the ANSYS problem that fits the studied cases of the present work has the following form:

$$\frac{\partial \sigma_{ij}^{A}}{\partial x_j} = 0 \tag{11}$$

and

$$\sigma_{ij}^{A}\Big|_{BS} = -\delta_{ij}\sigma_s \tag{12}$$

$$\left|v_i\big|_{SP}\right| = \delta \tag{13}$$

According to Eq. (12), the sintering stress can be interpreted as a hydrostatic compressive stress. Recalling, again, that the density and temperature gradients are negligible, the sintering sample can be considered to be homogeneous in each element. If a hydrostatic tensile stress is applied to a homogeneous body and that stress equals the sintering potential ($\sigma_x^{A}\big|_{BS} = \sigma_y^{A}\big|_{BS} = \sigma_z^{A}\big|_{BS} = -\sigma_s$), then contraction stops.[1] Hence, Eqs. (7) and (11)-(13) are physically equivalent to Eqs. (1) and (4)-(6) for calculating the stress state in a sintering body with the proper σ_s value.

This stress equilibrium problem (Eqs. (11)-(13)) with the isotropic form[1] of Eq. (7), the constitutive coefficients and the sintering stress of the Scherer Cell model[8] is the problem implemented in ANSYS (Fig. 1). Fig. 2 illustrates the sintering rectangle and its boundary conditions.

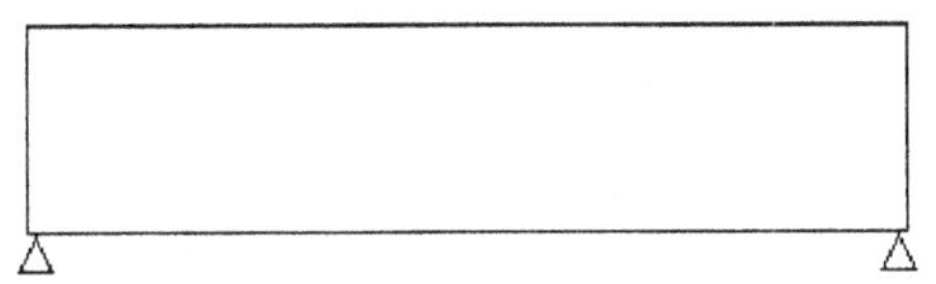

Fig. 2. The sintering rectangle is held by the two ends allowing a limited displacement.

RESULTS AND DISCUSSION

For calculations some parameters related to geometrical factors and materials properties are conveniently accommodated and the physical parameters herein are reported in relative units. The thermal distribution of the 2D rectangle in Fig. 3 shows a temperature difference of 100^{0}C along it. Fig. 4 illustrates the dependence of the relative density on the firing time for the temperature distribution shown in Fig. 3. Five points of this curve are pointed out. These points are referred to the five moments shown in Fig. 5. The left representations of this figure are the thermal stresses and the right side shows the shape evolution of the finite element model.

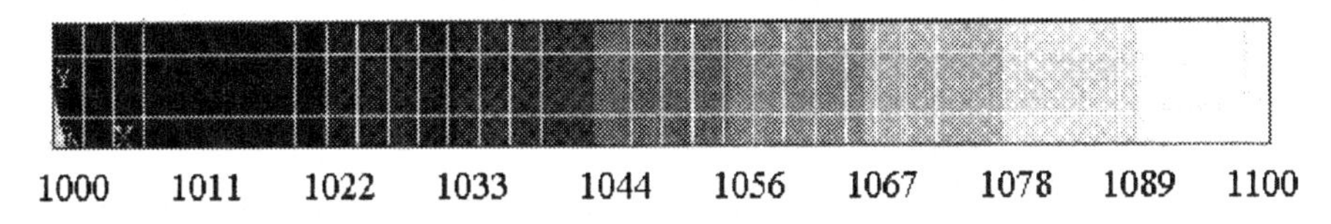

Fig. 3 Thermal profile (Temperature ^{0}C)

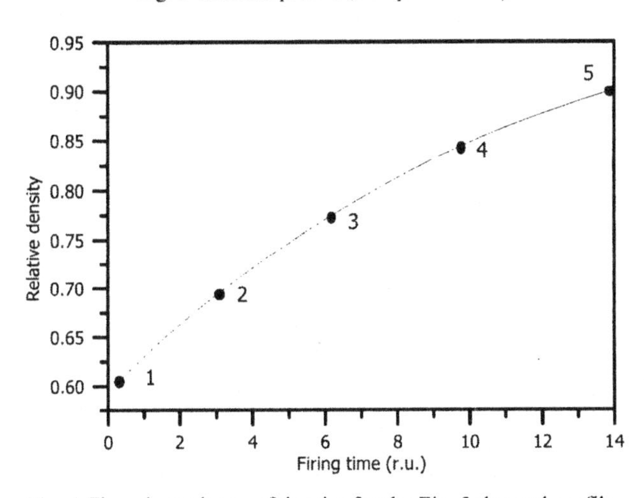

Fig. 4 Time dependence of density for the Fig. 3 thermal profile.

As can be noticed in Figs. 4 and 5, as densification advances, the curvature of the sintering 2D model increases. We studied several cases as the one shown in Figs. 4 and 5 considering different thermal and density gradients. Fig. 6 shows the dependence of the densification rate as a function of the firing time for several temperature and density gradients. It can be easily seen that the highest densification rate is reached for the total absence of gradients. This is an expected result because no retarding stresses must appear in the absence of gradients. Then, nearly free sintering is taking place. Fig. 6 also shows the dependence for a 10 times lower viscosity η. Consequently, the densification rate increases about 100 times.

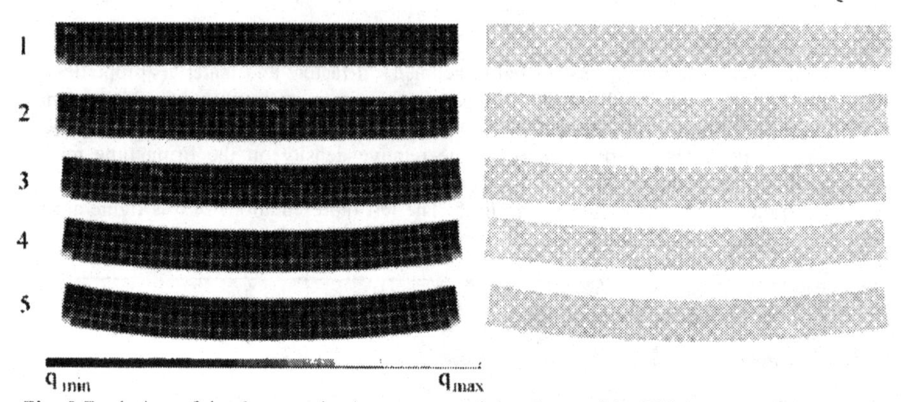

Fig. 5 Evolution of the thermo-elastic stresses and the shape of the FEM. q_{min} and q_{max} are the minimal and maximal von Mises stresses respectively. $q_{min} \times 10^6$ - $q_{max} \times 10^8$ have the values: (1) 0.38-0.14; (2) 0.33-0.31; (3) 0.24-0.5; (4) 0.74-0.71; (5) 0.34-0.79. q_{max} increases with time.

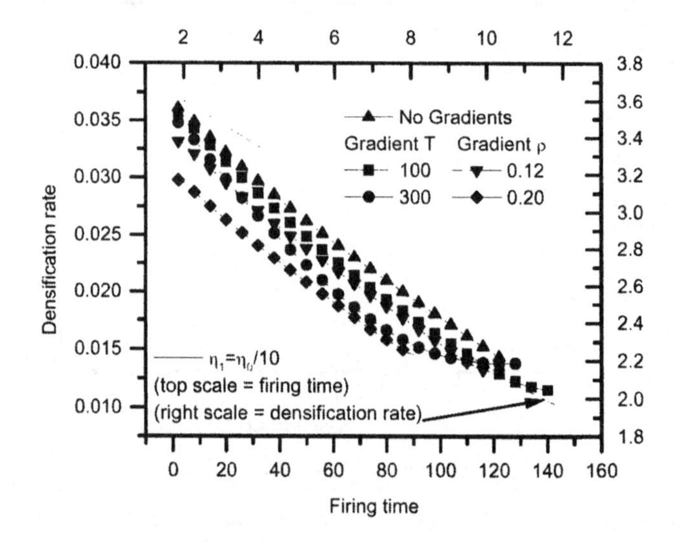

Fig. 6. Evolution of the densification rate as a function of the firing time for several temperature and density gradients. A case for a lower reference viscosity, η [8], is also shown.

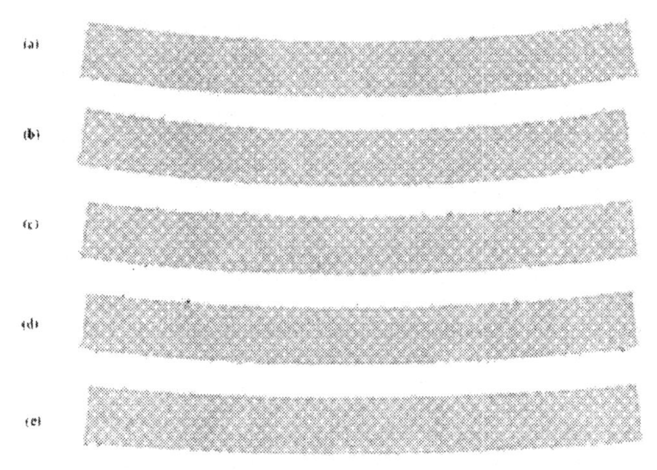

Fig. 7. $(a)\nabla T = 300, \nabla\rho = 0, h = 30; (b)\nabla T = 100, \nabla\rho = 0, h = 30; (c)\nabla T = 0, \nabla\rho = 0, h = 26;$ $(d)\nabla T = 0, \nabla\rho = 0.12, h = 22; (e)\nabla T = 0, \nabla\rho = 2, h = 19$. h is the difference between the highest (border) points and the lowest point of the top surface, this is a measure of surface curvature. According to the boundary conditions shown in Fig. 2 a curvature appears in all samples.

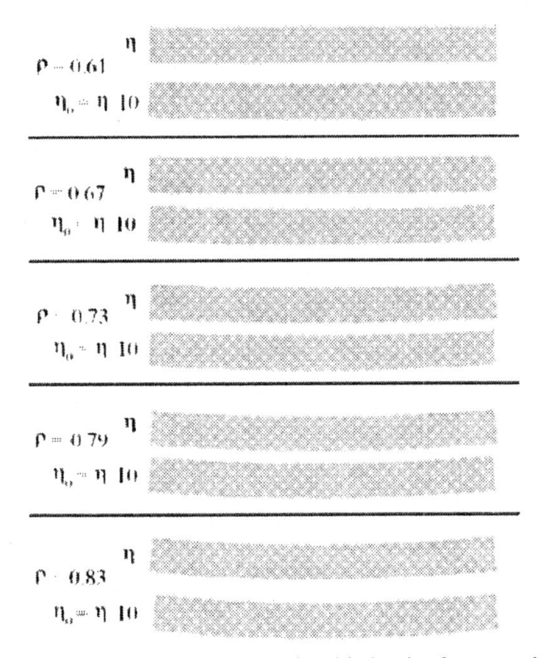

Fig. 8. Shape evolution of the sintering rectangle with density for two values of viscosity.

Fig. 7 shows the shape deformation for the cases: (a) Gradient T 300^0C, (b) Gradient T 100^0C, (c) No gradients, (d) Gradient ρ 0.12 and (e) Gradient ρ 0.20. For the case (d) and (e), it may be observed that deformation is minimal. However, we believe that the optimum firing conditions correspond to the case (c) since that this case has maximum densification rate with real minimum deformation. For the cases (d) and (e), green density has higher values due to the density gradients and sintering took less time. This is the real reason of less curvature for these cases. Fig. 8 illustrates that the less the viscosity value the higher the deformation during sintering.

CONCLUSIONS

A continuum approach to study densification and deformation during powder compact firing is proposed based on a viscous behavior during sintering. Thermal and density gradients are considered for this process. It may be concluded that the optimum processing conditions to minimize deformation and maximize densification are reached with the absence of thermal and density gradients. This case is similar to the free sintering conditions where no constrains are presented.

Special attention was focused on the viscosity parameter. It can be concluded that the smaller the viscosity, the higher the densification rate and the higher the shape deformation. Hence, in case the viscosity can be changed, for instance, with the help of a dopant, a sort of compromise must be established between densification and shape deformation for traditional sintering.

REFERENCES

[1]R. K. Bordia and G. W. Scherer, "On constrained Sintering- I. Constitutive Model for a sintering body. II. Comparison of Constitutive Models. III. Rigid Inclusions," *Acta Metall.*, **36** [9] 2393-2417 (1988).

[2]V. Cantavella Soler. "Simulación de la deformación de baldosas cerámicas durante la cocción," *PhD Thesis*, Castellón, 1998.

[3]D.-Z. Sun and H. Riedel. "Prediction of Shape distortions of hard metal parts by numerical simulation of pressing and sintering," *Simulation of Materials Processing: Theory, methods and Applications*, Shen & Dawson (eds) 1995 Balkema, Rotterdam, ISBN 90 54 10 553 4.

[4]T. Kraft, H. Riedel, O. Rosenfelder and P. Stingl. "Computational optimization of Parts Produced from Ceramic Powders," in: *Microstructures, Mechanical Properties and Processes*, Y. Bréchet (ed.), EUROMAT 1999, Vol. 3, Wiley-VCH, Weinheim, Germany (2000) 337-342.

[5]T. Kraft, O. Coube and H. Riedel, "Numerical Simulations of Pressing and Sintering in the Ceramic and hard Metal Industry," in: *Recent Developments in Computer Modelling of Powder Metallurgy Processes*, NATO Advanced research Workshop, Series III: Computer and Systems Science - Vol. 176, A. Zavaliangos, A. Laptev (eds.), IOS Press, Amsterdam, Netherlands (2001) 181 – 190.

[6]J Cedergren, NJ Sorensen, and A. Bergmark. "Three-dimensional analysis of compaction of metal powder," *Mechanics of Materials* **24** (2002) 43-59.

[7]H. Camacho, M.E. Fuentes, L. Fuentes, A. García and A. Pérez. "Evolución de la distribución de tensiones en un cuerpo cerámico durante la cocción. Parte 1: Planteamiento del problema," *Boletín de la Sociedad Española de Cerámica y Vidrio*, **42** [6] 353-359 (2003)

[8]G. W. Scherer, "Cell Models for Viscous Sintering," *Journal of the American Ceramic Society*, **74** [7] 1523-31 (1991).

APPLICATION OF A MICROSTRUCTURE-BASED MODEL FOR SINTERING AND CREEP

Markus Reiterer*, Torsten Kraft and Hermann Riedel
Fraunhofer-Institute for Mechanics of Materials
Wöhlerstr. 11
79108 Freiburg
Germany

ABSTRACT

A unified model for ceramic sintering and creep is presented. The model is based on the physical processes that are common to sintering and creep, *i.e.* grain boundary diffusion, grain boundary sliding, and grain coarsening. Additionally, the model incorporates interface reaction controlled diffusion, which is important for small grain sizes, and which leads to a nonlinear stress-strain rate relation. The model was applied to predict the sinter forming of reaction-bonded alumina. Some of the parameters required for the model were determined from sinter forging experiments, and these same experiments were used to test and validate the model predictions. Additionally, creep tests were conducted at three different temperatures, and the experimental results were compared with numerical predictions. Model predictions show good agreement with experimental sintering and creep results.

INTRODUCTION

Many ceramic materials have excellent properties that meet the demands of technical applications quite well. However, due to high processing costs ceramic materials are still not cost competitive in many applications. One challenge for the economical production of ceramic components is net-shape manufacturing. The sintered dimensions of a particulate ceramic body are determined by the shrinkage that occurs during densification. Variations in shrinkage can occur from differential sintering due to inhomogeneous green density distributions, temperature gradients, and gravity. Expensive mechanical finishing, *i.e.*, diamond grinding is the only means to eliminate unwanted dimensional deviations. If the densification-behavior of the green body during sintering is known, the shape of the green body can be designed to sinter to net-shape. This problem can be solved empirically, but it is both very time consuming and fairly complex. Computer simulations can provide a more efficient solution;[1] however, the accuracy of the computer solution does not always meet industrial requirements[2].

In this paper, we present a sintering model for sintering and creep. The sinter forming of reaction-bonded alumina (RBAO) is used to demonstrate the capabilities of a unified microstructure-based model for sintering and creep. Sinter forming is a process that combines pressure-enhanced sintering and plastic deformation[3] to produce advanced ceramic components. In contrast to hot pressing, where dense semi-finished parts are deformed, sinter forming starts with porous particulate bodies. (If sinter forging is performed in a closed die to achieve near-net-shaped parts, it is called sinter forming.) There are several advantages of sinter forming: 1) The deformation rate (starting with a porous body) is higher than of an already dense material; 2)

*Currently at: Sandia National Laboratories, MS 1349; Ceramic Materials Department, 1843; Advanced Materials Laboratory – Suite 100, 1001 University Blvd., S.E., Albuquerque, NM 87106

Nearly defect-free and fine-grained microstructures can be produced without liquid-phase sintering;[4] 3) Components with significantly improved mechanical strength can be produced;[5,6] and the quality of the finished surface can be controlled by the tooling.

When sinter forming is applied to a complex-shaped component, some regions can densify faster than others due to the different stress states[7]. Near the completion of sinter forming, when some regions have reached their final density, other areas of the part are still sintering. As a consequence, the denser regions can be deformed by creep. Therefore, a numerical model for sinter forming has to consider both, sintering and plastic deformation.

A MODEL FOR SINTERING AND CREEP

A microstructure-based model for solid state sintering, developed mainly by Riedel and Svoboda,[8-11] was chosen to simulate sinter forming. A detailed summary of the model is given in[12] or[13]. The concept of the model with regard to sintering mechanisms is based on the work of Ashby[14,15] and the mechanical aspects are based on the work of Jagota and Dawson[16] and McMeeking and Kuhn[17]. This model, as presented in[12] takes into account diffusive transport of matter and grain coarsening, and distinguishes between open and closed porosity. The constitutive equation is expressed as a relation between the macroscopic strain rate tensor and the stress tensor:

$$\dot{\varepsilon}_{ij} = \frac{\sigma'_{ij}}{2G} + \delta_{ij}\frac{\sigma_m - \sigma_s + \Delta p}{3K} \tag{1}$$

where σ'_{ij} is the stress deviator, σ_m is the hydrostatic stress, Δp is a gas pressure that can develop in closed pores, δ_{ij} is the Kronecker symbol, G and K are shear and bulk viscosity, respectively, and σ_s is the sinter stress that arises from the pore surface tension force.

In its original form, the model predicts a linear stress/strain-rate relation, i.e. K and G are independent of the stress. Various authors have reported a non-linear stress/strain-rate dependency for oxide ceramics when they are deformed at elevated temperatures and at stresses between 5 and 100 MPa.[18-21] Pan and Cocks[22] published a numerical model for final stage sintering that utilizes a quadratic stress/strain-rate relation. Creep is a factor in sinter forming; however, because creep is a special case of sintering under load, the same deformation mechanisms should be active. One reasonable explanation for this behaviour is that the grain boundaries do not act as perfect sources and sinks for vacancy diffusion. In some cases, it has been assumed that an interface reaction is necessary to the nucleate vacancies.[18-20] In metals analogy it is supposed that vacancies are generated or annihilated by grain boundary dislocation glide or climb. A theoretical analysis of the stresses on a moving grain boundary dislocation leads to the strain rate:[23]

$$\dot{\varepsilon}_{ir} = M\frac{\sigma^2}{r}, \tag{2}$$

where M is proportional to the mobility of the grain boundary dislocations, σ is the normal stress and r is the grain radius. Because vacancy nucleation and grain boundary dislocation motion are serial processes, the overall strain-rate becomes:

$$\frac{1}{\dot{\varepsilon}} = \frac{1}{\dot{\varepsilon}_d} + \frac{1}{\dot{\varepsilon}_{ir}}. \tag{3}$$

The stain-rate associated with pure Coble creep is:

$$\dot{\varepsilon}_d = C\frac{\sigma}{r^3}, \tag{4}$$

Inserting Eq. 2 and 4 into Eq. 3, we achieve

$$\frac{1}{\dot{\varepsilon}} = \frac{1}{C}\frac{r^3}{\sigma} + \frac{1}{M}\frac{r}{\sigma^2}. \tag{5}$$

Reinserting Eq. 4 and rearrangement leads to:

$$\frac{1}{\dot{\varepsilon}} = \frac{1}{\dot{\varepsilon}_d}\left(1 + \frac{\alpha}{\sigma r^2}\right) \tag{6}$$

where α is an adjustable parameter. Subsequently, the bulk (K) and shear viscosity (G) can be written as:

$$K = K_{lin}\left(1 + \frac{\alpha}{\overline{\sigma}\,\overline{R}^2}\right), \qquad G = G_{lin}\left(1 + \frac{\alpha}{\overline{\sigma}\,\overline{R}^2}\right) \tag{7}$$

where the linear viscosities, K_{lin} and G_{lin}, are obtained from reference[8] and the here derived term for the interface reaction-controlled diffusion, $\overline{R}$ is the mean grain radius, and $\overline{\sigma}$ is the effective stress given by:

$$\overline{\sigma} = \tfrac{1}{2}\left|\sigma_m - \sigma_s + \Delta p\right| + \tfrac{1}{2}\sigma_e, \tag{8}$$

where σ_e is the von Mises equivalent stress. By including interface reaction controlled diffusion, the model can consider any stress exponent n ($\dot{\varepsilon} \propto \sigma^n / r$) between 1 and 2. The term $\alpha / \overline{\sigma}\,\overline{R}^2$ shows that the interface reaction is more important for low stresses and small grain sizes.

To accurately predict creep, the existing sintering model was improved in three ways. First, at full density, where the material becomes viscous-incompressible, consequently, the bulk modulus becomes infinite. Therefore, the trace of the strain-rate tensor is assumed to be 0 when the porosity, f, is $< 10^{-6}$,

$$\dot{\varepsilon}_{kk} = \frac{\sigma_m - \sigma_s + \Delta p}{K} \equiv 0, \tag{9}$$

Second, at very low porosity, the gas pressure, Δp, in a closed pore increases and leads to an unrealistically high value of the effective stress, σ_e. To avoid this problem, when $f < 10^{-6}$, $\overline{\sigma} = \sigma_e$. Finally, an adjustable parameter was introduced to adjust the predicted strain rates within a small range.

SINTER FORGING AND CREEP EXPERIMENTS

For good formability, grain growth in the RBAO has to be suppressed. This can be accomplished with the addition of ZrO_2[6]. 20 vol% 2Y-ZTP was milled together with 44 vol% Al and 36 vol% Al_2O_3 powder in an organic liquid. This material will be referred to as the pure RBAO. The dried powder was uniaxially pressed and subsequently oxidized in air at temperatures between 350 and 1000 °C. To reduce the sinter forming temperature, 2% CuO and 2% TiO_2 were added to the zirconia modified RBAO[24]. This powder is referred to as the doped RBAO.

At the IWM in Freiburg sinter forming experiments were performed of cylindrical samples of pure and doped RBAO at different stresses and sintering temperatures to identify the material parameters for the model. Reaction bonded samples supplied by the TU Hamburg-Harburg (TUHH) were used for these experiments. The presintered specimens were placed in the sinter forging device (a loading dilatometer). Then, the furnace was heated to the sinter forging temperature at a rate of 10 K/min, and the samples were loaded with different axial stresses for 60 to 150 minutes. The axial load, σ_z, was applied within 300 s of reaching the maximum processing temperature. The axial force, specimen height, and specimen diameter were recorded continuously during the experiments. To characterize the plastic behaviour of the dense RBAO materials, step creep tests were also performed at the TUHH. Sinter forged specimens of each material were loaded with a rising stress from 10 to 100 MPa at three different temperatures.

Because grain coarsening has a significant influence on sintering and creep, grain size evolution was determined by means of an IMAGEC image analyses system. The grain diameters before sinter forging, after sinter forging, and after sinter forging and creep are given in Table I:

Table I: Evolution of grain size

		t in min	T in °C	D in µm
pure RBAO	before sinter forging	0	-	0.30
	after sinter forging	30*	1400	0.36
	after creep	32**	1450	0.83
doped RBAO	before sinter forging	0	-	0.33
	after sinter forging	30*	1150	0.88
	after creep	45**	1150	1.22

*time at sinter forging temperature
**time at creep temperature

DETERMINATION OF MATERIAL PARAMETERS AND SINTERING SIMULATIONS

The material parameters for modeling were determined using a PC-FORTRAN program to match the evolution of the density, strains, and grain size to the experimental results. Figure 1 shows the evolution of the relative density at a maximum temperature of 1450°C with axial stresses between 5 and 30 MPa. The axial and radial strains for the same conditions are plotted in Figure 2. Model predictions (dotted curves) of both densification and deformation are in good agreement with the experimental data. The activation energy and the pre-exponential factors for the three different diffusion mechanisms and the grain growth kinetics have the most important influence on the predicted results.

The parameters of both RBAO materials examined are presented in Table II. More detailed results are presented with a parameter study elsewhere.[25] The external gas pressure, initial density, and grain size were determined from experiments. The diffusion coefficients and grain boundary mobility were calculated, and are in good agreement with published work[18,21]. The remaining parameters were found in textbooks (molecular volume, surface energy and dihedral angle), or are model specific.

Table II: Parameters of the model for sinter forging of RBAO. (The parameters are defined *i.e.* in Ref.[9])

Parameter	Symbol	pure RBAO	doped RBAO
Initial relative density	ρ_0	0.59	0.615
Initial grain size	$\bar{R}_0$	0.15 μm	0.165 μm
External gas pressure	p_{ex}	0.1 MPa	0.1 MPa
Molecular volume	Ω	1.42e-29 m³	1.42e-29 m³
Grain boundary diffusion	δD_b^0	1.05e-9 m³/s	5.5e-9m³/s
	Q_b	475 000 J/mol	375 000 J/mol
Surface diffusion	δD_s^0	1.05e-7 m³/s	5.5e-7 m³/s
	Q_s	475 000 J/mol	375 000 J/mol
Volume diffusion	D_v^0	7.8e-2 m²/s	7.8e-2 m²/s
	Q_v	800 000 J/mol	800 000 J/mol
Grain boundary mobility	$\gamma_b M_b^0/4$	1.6 m²/s	3.75e+4 m²/s
	Q_m	572 000 J/mol	572 000 J/mol
Surface energy	γ_s	0.75 J/m²	0.75 J/mol
Dihedral angle	ψ	60°	60°
Initial derivation of the Hillert grain size distribution	δ	0.5	0.5
Pore detachment	β_0	1.0	1.0
G/K for open porosity	β_1	0.28	0.27
Multiplier for *G/K* at closed porosity	β_2	0.5	0.5
Multiplier for *G* at full density	β_3	0.3	0.4
Interface reaction parameter	α	8.25e-7 N	6.0e-6 N

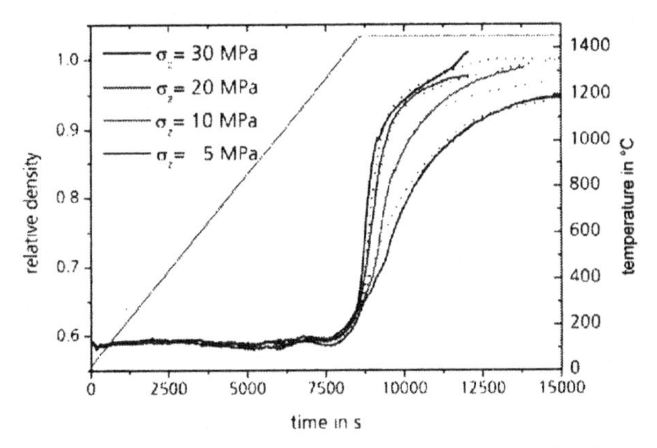

Fig. 1: Stress dependence of relative density during sinter forging tests preformed on pure RBAO. Solid lines: experimental results; Dotted curves: numerical predictions.

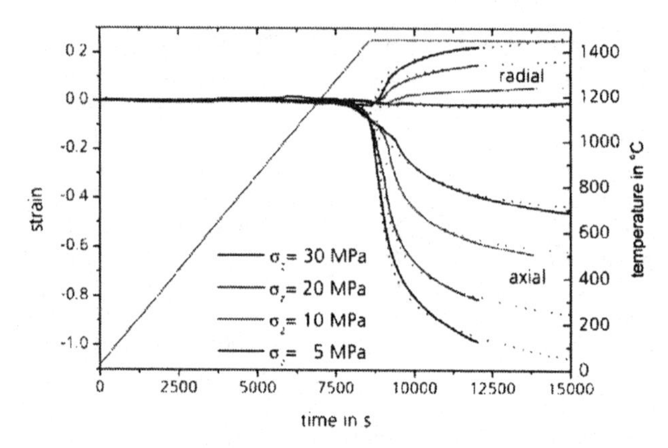

Fig. 2: Axial and radial strain during of sinter forging tests preformed on pure RBAO. Solid lines: experimental results; Dotted curves: numerical predictions.

SIMULATION OF THE CREEP EXPERIMENTS

Using the material parameters in Table II, a PC-FORTRAN program was used to predict the creep behaviour of the RBAO materials. For these calculations, sinter forging was simulated prior to creep. In Figure 3 strain-rate vs. stress plots are shown for both RBAO materials at three different dwell temperatures. The experimental data (squares) are compared to the numerical prediction (circles) and the sinter forging tests (triangles). Because density and axial stress change slightly during sinter forging, the creep rates are not constant (Fig. 3a & b). In Figure 3d, the minimum and the maximum creep rate for the three different experiments are shown. As expected, the simulation predicts results that are consistent with the experimental results from the sinter forging tests, as they were used to obtain the modeling parameters. An extrapolation to higher stresses should also yield accurate predictions, as the stress exponent n for the experimental results and the simulation is nearly identical. At a temperature of 1475°C (Fig. 3c) the experiment shows a decrease of the strain rate with increasing stress (time). The simulation shows the same trend. This effect is attributed to increased grain growth at higher temperatures.

DISCUSSION

For doped RBAO, the model predicted strain rates and stress exponents match the experimental results very well. In contrast, for the pure RBAO the simulated strain rates are 2 to 4 times higher than the test results. There are three possible reasons for this discrepancy: 1) The temperature of the creep experiments (TUHH) may have been higher than the sinter forging temperature (IWM); 2) The grain size of the samples used for the sinter forging tests at the IWM may have been larger than that in the samples tested at the TUHH; 3) The friction between the pressing punches and the samples may have differed. Both 1 and 2 have been assessed by simulation using the PC-FORTRAN program. A temperature difference of 50°C or a grain size difference of 33% could account for the discrepancy. The influence of friction during sinter forging was investigated using a finite element (FE) simulation. The model was implemented as a user-subroutine VUMAT in ABAQUS/Explicit®. Figure 4 shows the original and the deformed mesh of a cylindrical pure RBAO sample. A load of 20 MPa was applied for 60 min at 1450°C assume ing a friction coefficient of 0.15 in the presented case. The measured and calculated strains for the experiment are plotted in Figure 5. It is obvious that different friction coefficients in the simulation (0.1, 0.15 and 0.3) influence the axial strain and axial strain rate. A detailed study of the effect of friction coefficient is given in Ref.[25].

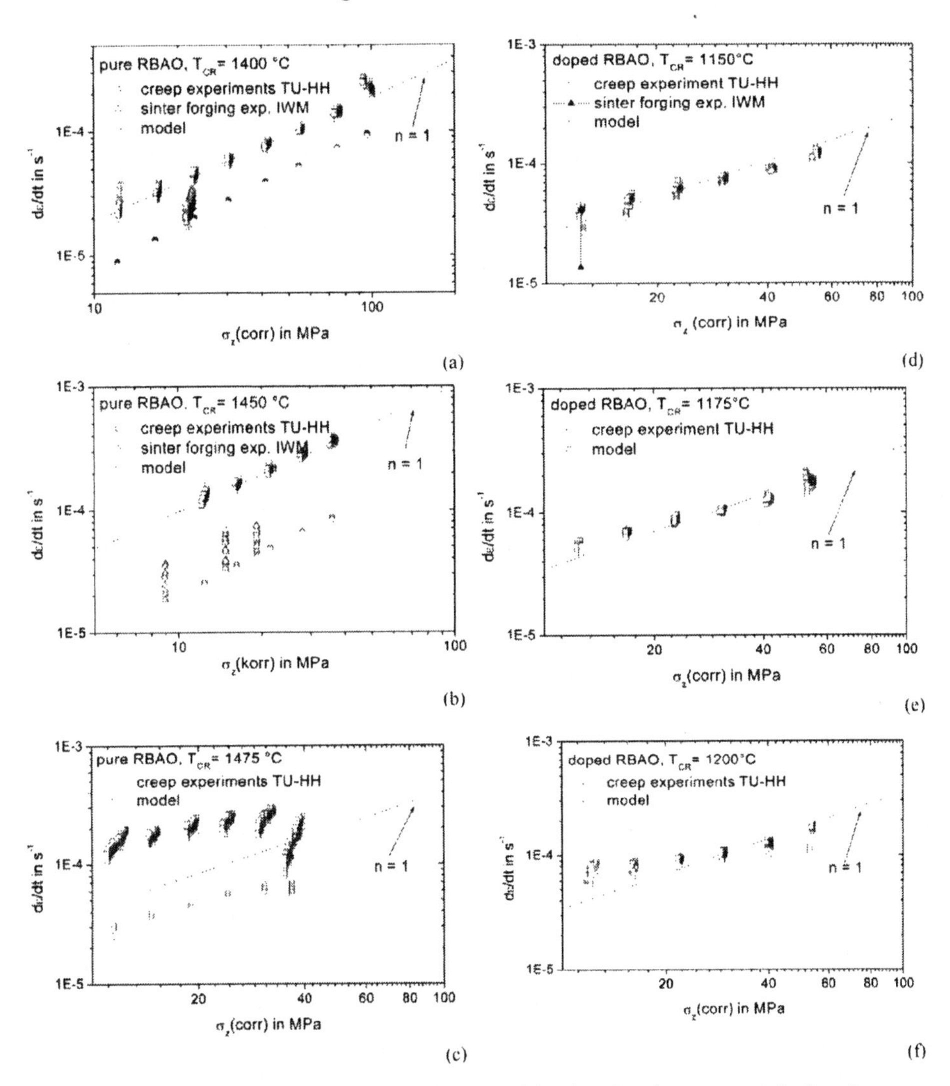

Fig. 3: Norton-plots of the creep experiments of the (nearly) dense material after sinter forging.

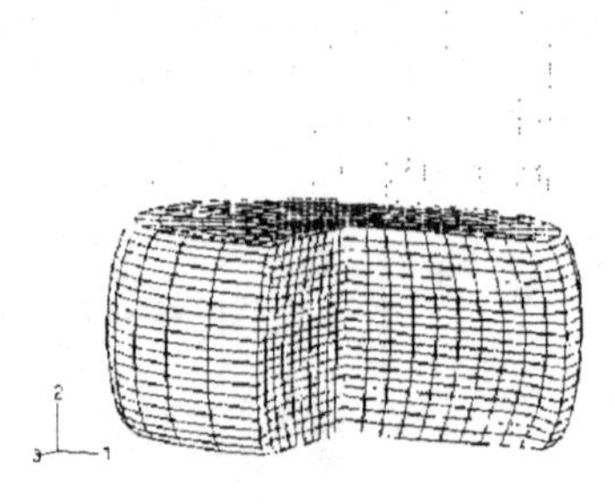

Fig. 4: Original and deformed mesh of a FE simulation of sinter forging a cylindrical sample

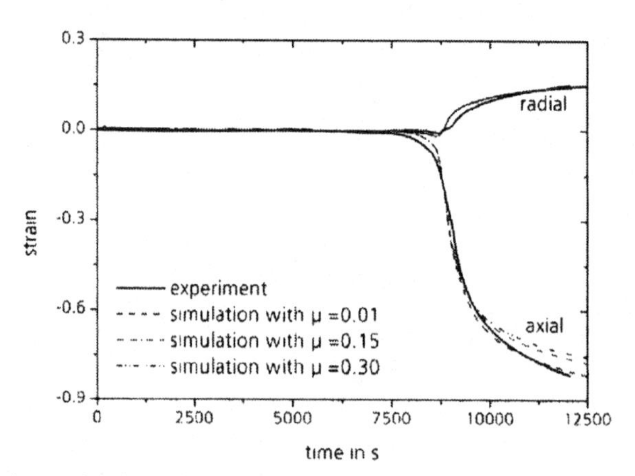

Fig. 5: Experimental and simulated strains with different friction coefficients.

Finally, the observed stress exponent, *n*, is about 1 for the pure RBAO and about 0.8 for the doped RBAO material. This was not expected, as an interface reaction generally results in a stress exponent greater than 1. However, because grain growth occurs concurrently, it can retard the deformation, and consequently, decreases the stress exponent.

SUMMARY

A model has been developed to predict ceramic sintering and creep. It has been implemented as user-subroutine in ABAQUS/Explicit®, and it was used to simulate among other geometries[7,25] the sinter forging of cylinders. The model has been applied to sinter forging, sinter forming, and creep. The developed methods enable one to predict model parameters and microstructure (density or grain size distribution). The introduction of interface reaction controlled creep leads to better results of sintering simulations, especially for very fine materials.

ACKKNOWLEDGEMENT

We wish to acknowledge financial support of Deutsche Forschungsgemeinschaft through grant number KR 1729/2-1 and the Advanced Ceramics Group at the TU Hamburg-Harburg for providing the samples and performing the creep tests.

Sandia National Laboratories is a multi-program laboratory operated by Sandia Corporation, a Lockheed Martin Company for the United States Department of Energy under contract No. DE-AC-94AL8500.

REFERENCES

[1]T. Kraft and H. Riedel, "Numerical Simulation of Die Compaction and Sintering," *Powder Metallurgy*, **45** 227-231 (2002).

[2]R.M German, "Critical Overview of Sintering Computer Simulations," International Conference on Powder Metallurgy & Particulate Materials, PM2Tec 2002, MPIF, Princeton, NJ 9-1 – 9-15 (2002).

[3]R. Janssen, R. Kauermann and N. Claussen, "Forming of Ceramics at Elevated Temperatures," In: *Towards Innovation in Superplasticity II*, Materials Science Forum, **304-306** 719-726 (1999).

[4]O. Kwon, C.S. Nordahl and G.L. Messing, "Submicrometer Transparent Alumina by Sinter Forging Seeded Gamma-Al_2O_3 Powders," *Journal of the American Ceramic Society*, **78** 491-494 (1995).

[5]K.R. Ventakachari and R. Raj, "Enhancement of Strength trough Sinter Forging," *Journal of the American Ceramic Society*, **70** 514-520 (1987).

[6]N. Claussen, R. Janssen and D. Holz, "Reaction Bonding of Aluminium Oxide." *Journal of the Ceramic Society of Japan*, **103** 749-758 (1995).

[7]M. Reiterer, T. Kraft and H. Riedel, „Manufacturing of a Gear Wheel Made from Reaction Bonded Alumina – numerical simulation of the sinter forming process," *Journal of the European Ceramic Society*, **24** 239-246 (2004).

[8]H. Riedel and J. Svoboda, "A Theoretical Study of Grain Coarsening in Porous Solids", *Acta Metall. Mater.*, **41** 1929-1936 (1993).

[9]J. Svoboda, H. Riedel and H. Zipse, "Equilibrium Pore Surfaces, Sintering Stresses and Constitutive Equations for the Intermediate and Late Stages of Sintering – Part I: Computation of Equilibrium Surfaces," *Acta Metall. Mater.*, **42** 435-443 (1994).

[10]H. Riedel, H. Zipse and J. Svobada, "Equilibrium Pore Surfaces, Sintering Stresses and Constitutive Equations for the Intermediate and Late Stages of Sintering – Part II: Diffusional Densification and Creep," *Acta Metall. Mater.*, **42** 445-452 (1994).

[11]H.Riedel, V. Kozák and J. Svoboda, "Densification and Creep in the Final Stage of Sintering," *Acta Metall. Mater.*, **42** 3093-3103 (1994).

[12]H. Riedel and B. Blug, "Comprehensive Model for Solid State Sintering and Its Application to Silicon-Carbide", pp 49-70; In: *Multiscale Deformation and Fracture in Materials and Structures: The J.R. Rice 60th Anniversary Volume*, Solid Mechanics and Its Application **84**, eds: T.J. Chuang and J.W. Rudnicki, Kluwer Academic Publishers, Dordrecht, (2001).

[13]T. Kraft and H. Riedel, "Numerical Simulation of Solid State Sintering-Model and Application," *Journal of the European Ceramic Society*, **24** 345-361(2004).

[14]M.F. Ashby, "A First Report on Sintering Diagrams," *Acta Metall.*, **22** 275-289 (1974).

[15]M.F. Ashby, "HIP 6.0 Background Reading," University of Cambridge, 1990.

[16]A. Jagota and P.R. Dawson, "Micromechanical Modelling of Powder Compacts - Unit Problems for Sintering and Traction Induced Deformation, *Acta Metall.*, **36** 2551-2561 and 2563-2573 (1988).
[17]R.M. McMeeking, and L.T. Kuhn, "A Diffusional Creep Law for Powder Compacts," *Acta Metall. Mater.*, **40** 961-969 (1992).
[18]R.M. Cannon, R.M., W.H. Rhodes and A.H. Heuer, "Plastic Deformation of Fine-Grained Alumina (Al_2O_3): I Interface-Contrelled Diffusional Creep," *Journal of the American Ceramic Society*, **63** 46-53 (1980).
[19]M.M.R. Boutz, C. von Minden, R. Janssen and N. Claussen, "Deformation Processing of Reaction Bonded Alumina Ceramics," *Materials Science and Engineering A*, **233** 155-166 (1997).
[20]A.H. Chokshi, "Diffusion Creep in Oxide Ceramics," *Journal of the European Ceramic Society*, **22** 2469-2478 (2002).
[21]O.A. Ruano, J. Wadsworth and O.D. Sherby, "Deformation of Fine-Grained Alumina by Grain Boundary Sliding Accommodated by Slip," *Acta Materialia*, **51** 3617-3634 (2003).
[22]J. Pan and A. C. F. Cocks, "A Constitutive Model for Stage 2 Sintering of Fine Grained Materials—II. Effects of an Interface Reaction," *Acta Metallurgica et Materialia*, **42** 1223-1230 (1994).
[23]B. Burton, "Interface Reaction-Controlled Diffusional Creep. A Consideration of Grain Boundary Dislocation Climb Sources," *Materials Science and Engineering*, **10** 9-14 (1972).
[24]W.R. Cannon, "High Ductility in Alumina Containing Compensating Additives," pp. 741-749; In: *Advances in Ceramics*, **10**, Structure and Properties of MgO and Al2O3 Ceramics, ed: W. D. Kingery, American Ceramic Society. Columbus, Ohio (1984).
[25]M. Reiterer, "Simulation of the Hot Forming of Reaction Bonded Alumina," Dissertation, University of Leoben (2004) in German.

Microstructure and Properties

PROCESSING AND MICROSTRUCTURE CHARACTERIZATION OF TRANSPARENT SPINEL MONOLITHS

Hans-Joachim Kleebe and Ivar E. Reimanis
Colorado School of Mines, Metallurgical and Material Engineering Department, Golden, CO 80401, U.S.A.

Ron L. Cook
TDA Research, Incorporated, Wheat Ridge, CO 80033, U.S.A.

ABSTRACT

Processing of transparent spinel monolithic discs was performed employing a unique powder processing route followed by hot-pressing of the powder pellets. $MgAl_2O_4$ powders were produced in a three step, single-pot process. The first step involves hydrolysis of Al-sec-butoxide ($Al(CH_3CH(O)CH_2CH_3)_3$ (ASB) to produce a boehmite sol. After hydrolysis was started, the mixture was peptized using a carboxylic acid. The boehmite powder was then converted to carboxylate-surface-modified boehmite nanoparticles. $Mg(acac)_2$ was added to the reaction mixture to exchange Al^{3+} at the boehmite surface. The spray-dried Mg-doped boehmite powders were then heated to 1200 °C to produce $MgAl_2O_4$ powders for use in the hot-pressing experiments. Hot-pressing was performed at a maximum temperature of 1550 °C for two hours with an applied pressure of 35 MPa. The resulting transparency depended strongly on the impurity level of the starting powder.

INTRODUCTION

Magnesium aluminate ($MgAl_2O_4$ or $MgO{\cdot}xAl_2O_3$), commonly termed spinel, is a fascinating material with many potentially desirable and 'tunable' properties. As a result, it already has a wide variety of niche applications in metallurgical, radiotechnical, chemical, and defense industries, ranging from refractory furnace materials to transparent armor. However, its current use in many applications derives from empirical knowledge developed through 'trial-and-error' approaches. This is particularly evident in the processing of *transparent* spinel, where mechanisms of densification and the corresponding connections to optical and mechanical properties are not understood as yet. Most of the current uses of *transparent* spinel are in military applications where cost is not as critical as it is in other market sectors; examples include windows and domes for severe environments and transparent armor [1-4]. Some potential future applications include laser host material, windows for lasers, optical heat exchangers, windows for radio frequency powder injectors, plasma diagnostic devices and chomatographs.

$MgAl_2O_4$ exhibits the spinel structure, a cubic unit cell with 56 atoms and a lattice parameter of a=8.1 Å, space group *Fd3m* (No. 227 of the International Tables). In *normal spinel*, the oxygen ions are arranged in a nominal face-centered cubic (fcc) sublattice, commonly with an oxygen parameter of 0.387 (0.375 = perfect fcc-arrangement). This small deviation from the perfect ordering effectively opens the tetrahedral lattice sites for the larger Mg^{2+} ions and simultaneously reduces the size of the octahedral sites, which then accommodate the smaller Al^{3+} ions [5]. The Mg^{2+} cations occupy 1/8 of the existing 64 tetrahedral sites, while the Al^{3+} cations occupy 1/2 of the 36 octahedral interstitial sites. The overall cation sublattice represents the structure of the

C15 Laves phase $MgCu_2$ [6]. In *inverse spinel*, such as $FeFe_2O_4$ (magnetite) or $MgFe_2O_4$, the tetrahedral sites are occupied by trivalent ions, while the octahedral sites are randomly filled by equal proportions of di- and trivalent cations. Between the two extremes of normal and inverse spinel, a varying degree of *cation disorder* can exist, in which A- and B-site cations gradually exchange their position [5-11]. At 2000 °C, the solid solution in spinel ranges from about 40% to >80%, as shown in the corresponding phase diagram in Figure 1 [6,8].

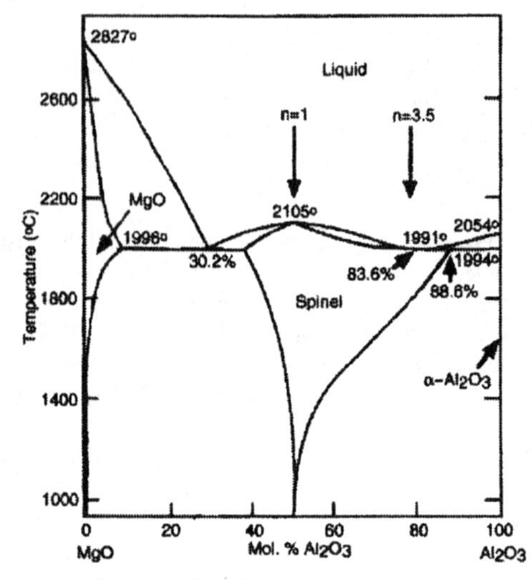

Fig. 1: MgO-Al_2O_3 phase diagram (in [6], adopted from Hallstedt [8]).

Fabrication of dense polycrystalline ceramics, starting from fine-grained powders, involves powder conditioning, powder packing, and densification, the latter of which can occur through various stages. It is the *latest stage of densification*, however, that is critical in determining the resulting optical transmissivity, since this stage determines the residual porosity that affects optical transparency. The final location and distribution of sintering aids, typically added to ceramics to aid in densification and impede grain growth, can also contribute to light scattering effects. Various studies have shown that relatively small variations in amounts and types of sintering additives may have large effects on the densification of spinel. It is generally accepted that even small amounts of additives, typically 0.5–1.0 wt.% of LiF, are needed to produce dense transparent spinel [2-4]. However, very little is known about the specific role of these additives. There are experimental data suggesting that their presence results in MgO-deficient spinel ($x>1$), which implies an increase in the oxygen vacancy concentration and hence an increase in sintering rate. Huang *et al.* [12] suggest a mechanism of enhanced densi-fication with LiF and LiF/$CaCO_3$ mixtures via the formation of a liquid or a vapor phase.

Spinel has a cubic structure and is therefore optically isotropic; thus, complex polycrystalline components can be fabricated without the severe scattering problems that are inherent to non-cubic crystal structures such as alumina. In the microwave region, the isotropy of spinel prevents localized absorption and heating which commonly occurs in non-cubic systems. Moreover, since spinel does not undergo any polymorphic phase changes upon cooling, changes in optical properties will only be a consequence of the densification mechanism. Spinel appears to be the preferred optical material for future seeker and sensor applications on missiles and aircrafts due to its better transmission characteristics in the mid infra-red and its index of refraction. Spinel offers a higher transmissivity than either ALON or sapphire in the mid infra-red region leading to improved signal-to-noise ratios with improved acquisition and tracking capabilities. In

Table 1, the absorption coefficient of spinel is compared with ALON and sapphire. The absorption coefficient of spinel (in the mid IR-band) shows a higher absorption edge around 5.5 μm (absorption coefficient of 1.3 at 500 °C) compared to either ALON or sapphire (with 3.7 and 2.1 at 500 °C, respectively). These characteristics make spinel very attractive for many high-temperature dome and window applications. In addition, sapphire suffers from birefringence (hexagonal structure) and ALON has far inferior transmission in the mid infra-red.

Material	25 °C	250 °C	500 °C
Spinel	0.4	0.7	1.3
Sapphire	0.8	1.3	2.1
ALON	1.6	2.4	3.7

Tab. 1 Absorption coefficient [cm^{-1}] of spinel, sapphire, and ALON at 5.0 μm; data shown were taken at room temperature, 250 °C, and at 500 °C.

EXPERIMENTAL PROCEDURES

The $MgAl_2O_4$ powders used to prepare transparent spinel parts were produced in a three step, single pot process. The first step was the hydrolysis of aluminum sec-butoxide ($Al(CH_3CH(O)CH_2CH_3)_3$ (ASB, Chattem Chemicals) to produce a boehmite sol. ASB from Chattem Chemicals is a liquid at room temperature and has less than 5ppm impurities. Hydrolysis of the ASB was carried out in 80°C distilled water. At 80 °C, the kinetically stable hydrolysis product is boehmite (i.e., Al(O)OH)) instead of $Al(OH)_3$ that forms by hydrolysis of ASB at lower temperatures. After hydrolysis was started the mixture was peptized using a carboxylic acid. Additional acid was then added to convert the nanosized boehmite particles to a carboxylate-surface-modified boehmite nanoparticles. The mixture was heated overnight and then $Mg(acac)_2$ was added to the reaction mixture along with additional water. The reaction mixture was then stirred for an additional two hours at 80°C wherein the magnesium from the $Mg(acac)_2$ exchanges with aluminum at the boehmite surface. The exchanged Al^{3+} forms aluminum acetylacetonate. The resulting mixture was then cooled to room temperature and spray-dried. This process produces a fine white powder with a high surface area of ~250 m^2/g and is indistinguishable from boehmite by XRD analysis (Figure 3).

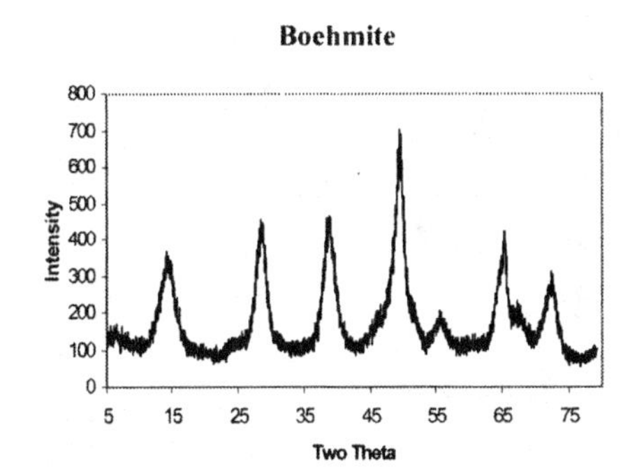

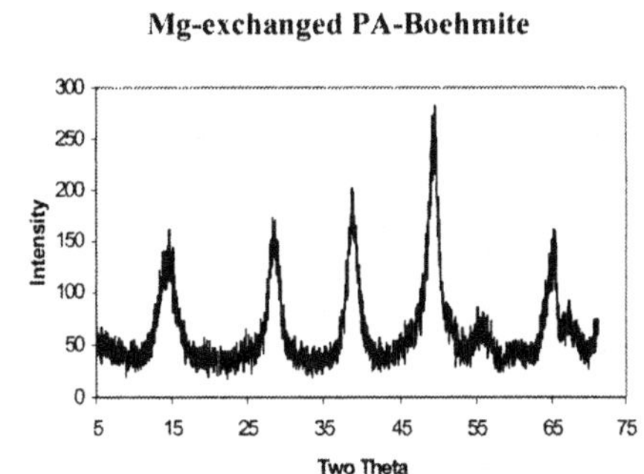

Fig. 3: XRD patterns for boehmite (left) and Mg-exchanged proprionic acid (PA) boehmite (right); both spectra were recorded under identical conditions.

The spray-dried Mg-doped boehmite powders were then heated to produce the $MgAl_2O_4$ powders for use in hot-press experiments. Heating the carboxylate-surface-modified boehmite powders to temperatures exceeding 500 °C causes the loss of the carboxylic acid at ~300 °C and transformation of the Mg-doped boehmite into Mg-doped γ-alumina by loss of water (e.g., $2Al(O)OH \rightarrow Al_2O_3 + H_2O$) at temperatures above 500 °C. No significant additional transformation of the Mg-doped Al_2O_3 after 900 °C is observed by XRD analysis. Further heating of the powders to above 900 °C in air results in the complete transformation of the Mg-doped alumina powders to $MgAl_2O_4$. At 1000 °C, the powders have a relatively high surface area of ~77 m^2/g, while additional heating to 1200 °C reduces the surface area to 24 m^2/g. After mixing with LiF (ranging from 0.0 to 1.0 wt%), the powders were densified using a Thermal Technology Inc. (model number 610G-25T) hot press. A graphite die (diameter 25.4 mm) fixture was used with graphite foil as a liner. A pre-load of 3.5 MPa was applied prior to heating. The specimens were heated at about 2 °C/min to 1550 °C and held for 2 hrs. A uniaxial die pressure of 35 MPa was applied at about 1550 °C. During the entire run a vacuum better than 10^{-5} Torr, was maintained. Microstructure characterization of the hot-pressed spinel samples was performed by scanning electron microscopy (SEM), using a FEI Quanta600 instrument operating at 30 kV, and by transmission electron microscopy (TEM) employing a FEI CM200STEM instrument operating at 200 kV. In order to minimize electrostatic charging under the electron beam, samples were gold-coated (SEM) and lightly coated with carbon (TEM studies).

RESULTS AND DISCUSSION

Dense polycrystalline spinel may be fabricated either by starting from powder mixtures of Al_2O_3 and MgO which upon heating react to form spinel or directly from $MgAl_2O_4$ powders. The former was used more frequently in the past, but is difficult to control and typically results in impurities that were introduced during the formation of the starting powders. During the last decade, substantial improvements have been made in the preparation of spinel powders [3-6,12], which has led to better control over purity, stoichiometry, starting particle size, and particle-size distribution. A method recently devised and patented by TDA Incorporated was employed here, allowing control over purity, particle size, composition, particle-size distribution, and surface area, as described in the Experimental Procedures. Calcination at relatively low temperatures (1000-1200 °C) results in the formation of a high-purity, fine-grained $MgAl_2O_4$ starting powder (Figure 4).

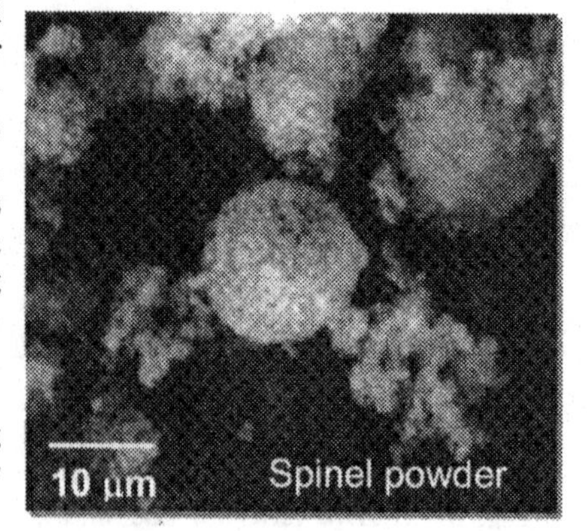

Fig. 4: SEM image of the $MgAl_2O_4$ starting powder upon calcination at 1200 °C (partly agglomerated).

Commonly, high pressures have been applied during processing to increase the infrared transparency to an acceptable range for optical applications. It seems likely that some plastic flow may be involved in densification of spinel powder compacts processed under high pressure. Thus, it is important to understand whether or not dislocations have a role in final densification. Preliminary TEM analysis of the different spinel samples fabricated showed a small number of spinel grains with dislocation networks. However, since those defect structures seemed to be less prominent in the LiF-doped samples investigated (0.75 and 1.0 wt%), it was assumed that they did not strongly affect the optical properties of the spinel samples processed here. Vapor-phase transport may be involved in an early stage of densification, and this may also influence the later stages of sintering. It is interesting to note that the activation energy for formation of spinel from MgO and Al_2O_3 is lowered by additions of LiF and combinations of LiF/$CaCO_3$ [12]. Thus, it may be that the sintering aids for spinel play multiple, interacting roles, with a dependence on whether or not starting powders comprise MgO and Al_2O_3 or $MgAl_2O_4$. The effect of sintering aid on the transmissivity is somewhat difficult to predict because LiF aids in densification, which is necessary for high transmissivity, but at the same time likely segregates at grain boundaries (no secondary phase formation) and scatters light. Specimens fabricated in an identical manner (densified in a graphite fixture vacuum hot press at 1550 °C, 3.5 MPa) but with slightly different amounts of LiF (0, 0.5 and 0.75 wt%) show significantly different light transmission data, as given in Figure 5. The starting powders were all calcined at 1200 °C prior to mixing with LiF.

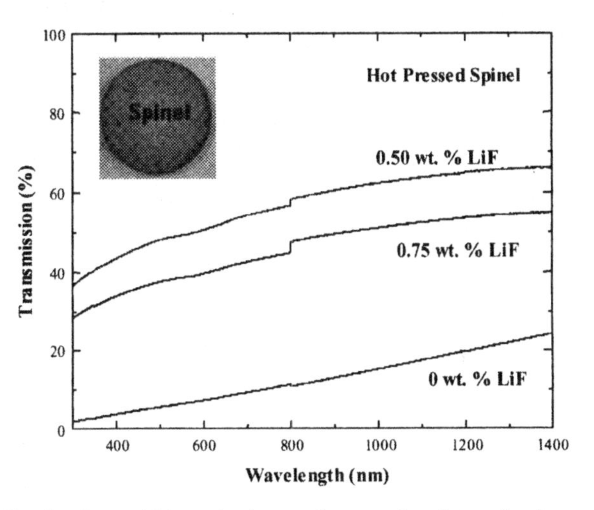

Fig. 5: Transmissivity from 300 nm to 1.4 μm of three dense spinel specimens. The specimen containing lower LiF content (see inset) shows a higher transmissivity across the entire range of light examined. The discontinuity at 800 nm is caused by a change in grating.

The final grain size and distribution within spinel samples can be determined on surfaces prepared either by chemical etch (phosphoric acid at 185 °C for 4 min) or thermal etch (1300 °C for 30 min). One example of a thermally etched microstructure of the hot-pressed sample doped with 0.75 wt% LiF is given in Figure 6. An important feature of this image is the large variation in grain size, a characteristic feature of all LiF-containing materials processed; however, its origin is not understood yet. Moreover, this sample revealed small precipitates along grain boundaries (Figure 6), which were identified as Zn-spinel. The initial boehmite powder used to process the spinel starting powder contained a minor amount of zinc. Hence, the purity level of all ingredients used to prepare dense spinel samples strongly effects the final product and in turn its optical transmissivity, which also explains why the overall transmissivity of the spinel samples prepared here is not higher (compare also Figure 5).

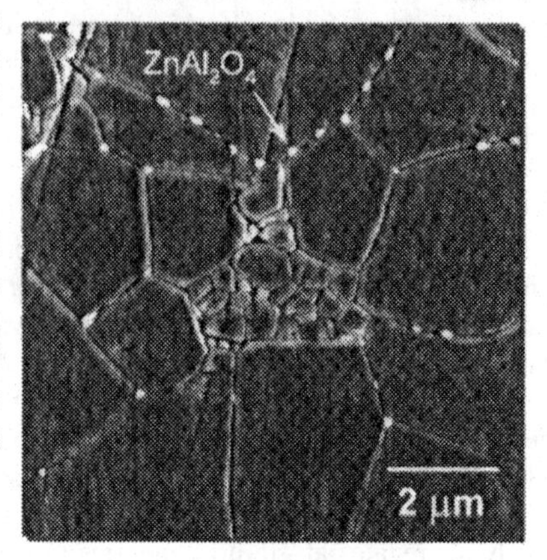

Fig. 6: SEM image of a thermally etched (1300 °C, 1h, air) spinel surface (0.75 wt.% LiF). Note the large variation in grain size (small cluster in the center of the image) and the small precipitates along grain boundaries ($ZnAl_2O_4$ impurities from the starting powder).

The $ZnAl_2O_4$ decorating the grain boundaries was found to be due to impurities in the $Mg(acac)_2$ precursor. However, after producing a high-purity $Mg(acac)_2$ precursor and producing of a higher purity powder, we were able to process a high-quality spinel part with outstanding optical properties, as shown in Figure 7. Though not shown in Figure 5, the transmissivity of this specimen exceeded 85 % over the majority of the wave-lengths.

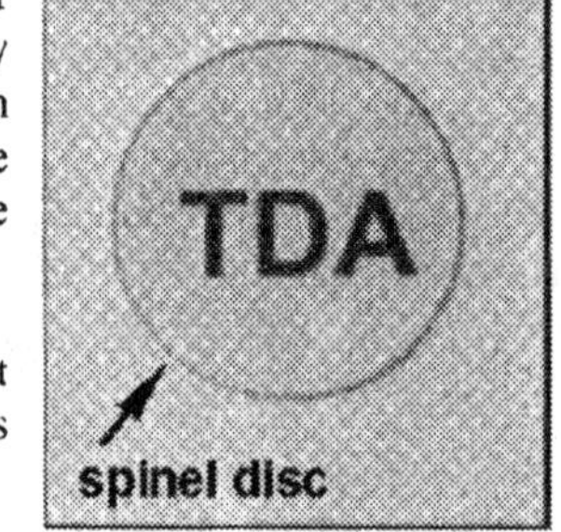

Fig. 7: Optical micrograph of the highly transparent spinel disc after the Zn-impurity in the Mg(acac)2 was removed.

Conventional and high-resolution TEM (HRTEM) was employed to study the atomic structure of internal interfaces. In particular, the question was addressed whether or not an amorphous grain boundary phase was present. In the case that LiF simply segregates at internal interfaces, the inner mean potential across grain boundaries will change depending on the amount of F-ions present and defocus Fresnel fringe imaging would enable the determination of LiF segregation. Earlier studies by White and Kelkar [16] have shown that LiF can exist as a segregant in isolated regions, rather than a continuous amorphous film, as would be expected for an intergranular glassy phase along grain boundaries. This result is consistent with our first preliminary observations using HRTEM imaging of grain boundaries of LiF-doped spinel. As can be seen in Figure 8, no clear evidence of a continuous amorphous film along the interface was observed, although faceting of grain boundaries commonly does not allow for a conclusive interpretation of the observed image contrast. It should be noted though that the presented result is representative for this material since seven interfaces were investigated by HRTEM, none of which revealed a continuous intergranular film. Similar results were obtained from the 1.0 wt% LiF-doped spinel sample.

Chiang and Kingery [14,15] performed energy dispersive X-ray analysis (EDX), operating in the windowless mode, using a scanning TEM to characterize the chemical composition of grain boundaries in spinels of different overall composition. Their findings suggest that strong deviations in stoichiometry are present at grain boundaries in *all* compositions of $MgAl_2O_4$, i.e., in stoichiometric and non-stoichiometric spinels. The Al/Mg ratio increased sharply across the grain boundary with an excess of Al relative to Mg per unit boundary area, which was equivalent to 0.5 to 2.0 monolayers, depending on the specific spinel lattice stoichiometry. However, it was noted that the excess Al amount was much less in Mg-rich compositions, as compared to stoichiometric or Al-rich spinel. Segregation of impurity cations such as Si and Ca at grain boundaries was found to be in general rather low (less than 0.2 monolayers), but varied from boundary to boundary. No systematic variation in impurity segregation with spinel stoichiometry was observed, and it was therefore concluded that the variations observed in grain-boundary mobility with spinel stoichiometry were not related to segregation of impurities.

Initial EDX analysis on the 0.75 and 1.0 wt% LiF-doped spinel samples was performed to verify as to whether grain-boundary segregation of LiF had occurred. No distinct F-signal could be detected, independent of the LiF content. However, it is possible that the corresponding (Li)F content is still below the intrinsic detection limit of the EDX analysis technique applied (about 1.0 wt%).

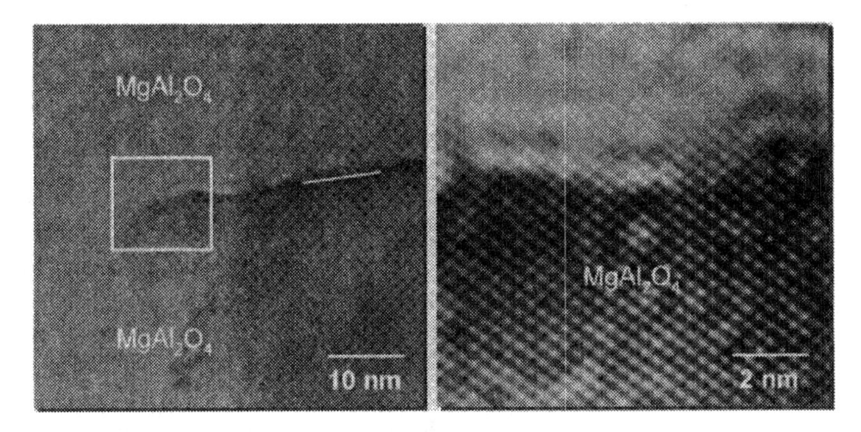

Fig. 8: HRTEM image of an interface in 0.75 wt.% LiF doped transparent spinel. The right side shows a Fourier-filtered image of the boxed-region on the left. Note that no clear indication of an amorphous intergranular film along the grain boundary could be detected.

CONCLUSIONS

Densification of the Mg-exchanged boehmite nanopowder, which was converted to nanosized spinel powder upon calcination at 1200 °C, via hot pressing at 1550 °C for 2 h under 35 MPa resulted in transparent spinel discs. Most samples prepared showed a slightly grayish color, limiting their overall transmissivity. A strong dependence of transmissivity in the infra-red with LiF content was observed. It is concluded that an optimum amount of LiF exixts (less than 0.75 wt%) at which the transmissivity is highest. EDX analysis of grain boundaries did not reveal a pronounced segregation of LiF to the interfaces, however, small grains were observed at grain boundaries (on thermally etched samples), which were identified as small Zn-spinel precipitates; a results of a minor Zn-impurity in the initial Mg-exchanged boehmite powder. Once this impurity in the $Mg(acac)_2$ precursor contamination was removed, transparent spinel with outstanding optical properties was produced. It is therefore concluded that that processing of transparent spinel is extremely sensitive to small impurity levels of the starting materials.

REFERENCES

1. E. A. Maguire, Jr. and R. L. Gentilman, "Press Forging Small Domes of Spinel," Ceramic Bulletin, **60** [2] 255-56 (1981).
2. J.-G. Li, T. Ikegami, J.-H. Lee, and T. Mori, "Fabrication of Translucent Magnesium Aluminate Spinel Ceramics," J. Am. Ceram. Soc., **83** [11] 2866-68 (2000).
3. D. W. Roy and J. L. Hastert, "Polycrystalline $MgAl_2O_4$ Spinel for High Temperature Windows," Ceram. Eng. Sci. Proc., **4** [7-8] 502-509 (1983).
4. R. J. Bratton, "Initial Sintering Kinetics of $MgAl_2O_4$," J. Am. Ceram. Soc., **52** [8] 417-19 (1969).
5. K. E. Sickafus, J. M. Wills, and N. W. Grimes, "Structure of Spinel," J. Am. Ceram. Soc., **82** [12] 3279-92 (1999).
6. T. E. Mitchell, "Dislocations and Mechanical Properties of $MgO-Al_2O_3$ Spinel Single Crystals," J. Am. Ceram. Soc., **82** [12] 3305-16 (1999).
7. F. S. Galasso, "Structure and Properties of Inorganic Solids," Pergamon Press, Oxford, U.K., (1970).
8. B. Hallstedt, "Thermodynamic Assessment of the System $MgO-Al_2O_3$," J. Am. Ceram. Soc., **75** [6] 1497-507 (1992).
9. S. P. Chen, M. Yan, J. D. Gale, R. W. Grimes, R. Devanathan, K. E. Sickafus, and M. Nastasi, "Atomistic Study of Defects, Metastable and 'Amorphous' Structures of $MgAl_2O_4$," Philos. Mag. Lett., **73** [2] 51-62 (1996).
10. C.-J. Ting and H.-Y. Lu, "Defect Reactions and the Controlling Mechanism in the Sintering of Magnesium Aluminate Spinel," J. Am. Ceram. Soc., **82** [4] 841-48 (1999).
11. T. Shiono, H. Ishitomi, Y. Okamoto and T. Nishida, "Deformation Mechanism of Fine-Grained Magnesium Aluminate Spinel Prepared Using an Alkoxide Precursor," J. Am. Ceram. Soc., **83** [3] 645-47 (2000).
12. J.-L. Huang, S.-Y. Sun, and Y.-C. Ko, "Investigation of High-Alumina Spinel: Effect of LiF and $CaCO_3$ Addition," J. Am. Ceram. Soc., **80** [12] 3237-41(1997).
13. C.-J. Ting and H.-Y. Lu, "Deterioration in the Final-Stage Sintering of Magnesium Aluminate Spinel," J. Am. Ceram. Soc. 83[7] 1592-98 (2000).
14. Y. M. Chiang and W. D. Kingery, "Grain Boundary Migration in Non-Stoichiometric Solid Solutions of Magnesium Aluminate Spinel: I, Grain Growth Studies," J. Am. Ceram. Soc., **72** [2] 271-77 (1989).
15. Y. M. Chiang and W. D. Kingery, "Grain Boundary Migration in Non-Stoichiometric Solid Solutions of Magnesium Aluminate Spinel: II, Effects of Grain-Boundary Non-stoichiometry," J. Am. Ceram. Soc., **73** [5] 1153-1158 (1990).
16. K. W. White and G. P. Kelkar, "Fracture Mechanisms of a Coarse-Grained Transparent $MgAl_2O_4$ at Elevated Temperatures," J. Am. Ceram. Soc., **75** [12] 3440-44 (1992).

PARTICLE BONDING TECHNOLOGY FOR COMPOSITE MATERIALS -MICROSTRUCTURE CONTROL AND ITS CHARACTERIZATION

Makio Naito and Hiroya Abe
Joining and Welding Research Institute,
Osaka University,
Mihogaoka 11-1, Ibaraki, Osaka 576-0047 Japan

ABSTRACT

Particle bonding technology is a promising approach to present designs of composite particles as well as nano/micro structural controls of composite ceramic parts. By making use of the unique properties of nanoparticle surface, the bonding can be well conducted at lower temperature without any binder in dry phase. Based on this concept, new apparatus to make various kinds of composite particles was developed. Novel composite particles were successfully fabricated, and the materials made from the composite particles showed interesting properties resulted from their nano/micro structural nature.

INTRODUCTION

Particle design and particle assembly control are of great interest in advanced ceramics, especially from the viewpoint of Nanotechnology. A particle with nano scale or materials with nano scale structures may exhibit new properties in comparison with larger scale structures, which could be used for new applications, in electronics, biochemical or medical field including drag delivery system. In the research of nano scaled particles and assemblies, fabrications of nano sized powders as well as the control of the interface structure and composition is regarded to be a key technology.

In this paper, a new particle bonding technology is introduced for making novel composite particles. This technology has mainly the following two features: At first, it achieves direct bonding between particles at lower temperature conditions in dry phase. By making use of this feature, design of composite particles is successfully conducted. The second feature is that this technology aims at nano/micro structural control by assembling of composite particles. As a result, it can control various kinds of nano/micro structure and can produce new materials in a more simple manufacturing process. Based on this concept, new apparatus to make composite particles was developed. In this paper, the apparatus will be introduced, and its application for new materials will be explained.

PROCESSING METHOD

This technology is based on mechanical method. The mechanical method utilizes mechanical energy instead of thermal energy to provide the activation for solid-state reaction. It can be applied as an activation process to promote reactions or reaction process by making use of the high temperature developed on the interface between particles during operation [1,2].

Figure 1 shows the schematic illustration of the device which has been recently developed to produce mechanical milling modes with glow discharge. The configuration of this device is based on Mechanofusion system (Hosokawa Micron Corp., Osaka, Japan) [3]. A micrograph of the device under operation with glow discharge is shown in Figure 2. The main parts are a rotating chamber and an arm head fixed with a certain clearance against the inside wall of the chamber.

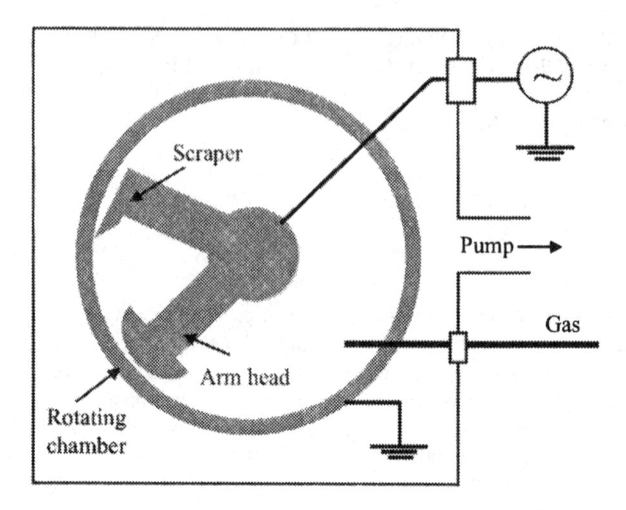

Figure 1, An illustration of the device with mechanical and electrical activation for particle-particle bonding

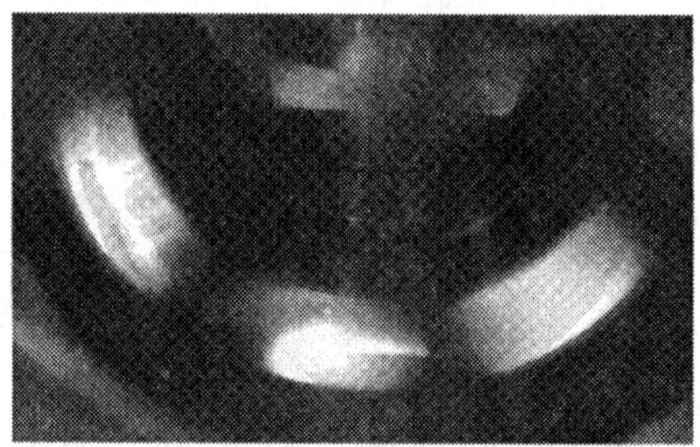

Figure 2, A micrograph of the device under operation. Chamber is rotating with applying a.c. electric power under Ar gas. The armhead and the scraper become visible by internal plasma.

The diameter of the chamber is 35 cm, and can be rotated up to 3500rpm. The arm head is fixed to the shaft which is not rotating. It is connected through the shaft to an outer electric power source operating with low frequency (60 Hz), as shown in Fig.1. On the other hand, the chamber is connected through the rotating shaft fixed to the bottom of the chamber to ground level. A gas line is piped into the chamber.

When the chamber rotating, powder material is compressed into the clearance and receives various kinds of mechanical forces such as compression, attrition and shearing. When applying the electric power into clearance simultaneously, the powder material will be mechanically activated in the state of non-equilibrium plasma. After the compression, the powder is dispersed by the scraper. These actions are repeated during the chamber rotation. It is noted that the rotating chamber has an advantage to prevent a concentration of discharge current into powder materials. The specific feature of the device is that no media ball is employed, which presents a highly beneficial effect with regard to contamination. It is also remarkable that the surface of particles can be effectively activated through particle-particle surface friction and plasma ambient.

EXPERIMENTAL RESULTS AND DISCUSSION

I. PARTICLE DESIGN BY MECHANICAL METHOD

Prior to the operation with glow discharge, it demonstrates that the mechanical method can be used to make novel composite particles. Figure 3 shows an example where silica particles (2 μm) are successfully coated with nanosized TiO_2 (15nm) by the mechanical method. XPS analysis of the particles revealed that the $Ti_{2}p$ peak of binding energy of composite particles has shifted its peak from the initial state [3]. The results obtained indicate that a chemical interaction has occurred on the interfaces between the SiO_2 core particles and TiO_2 nanoparticles by using the apparatus.

Figure 4 shows another example of making composite particles in a similar manner. In this case, resin particles of fairly uniform size (100 nm) were coated on the surface of spherical particles. It is just like closed packing structure of particles on the solid surface. By making use of mechanical action such as friction and rolling between particles, highly ordered particle structure can be successfully formed.

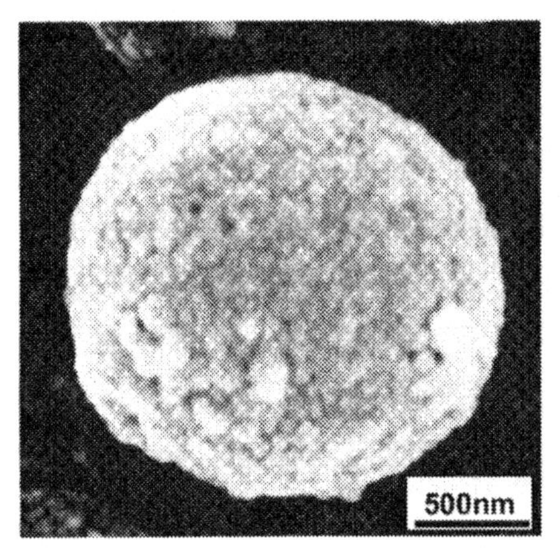

Figure 3, Silica particle (2μm) coated with nanosized TiO_2 (15nm) by the mechanical method.

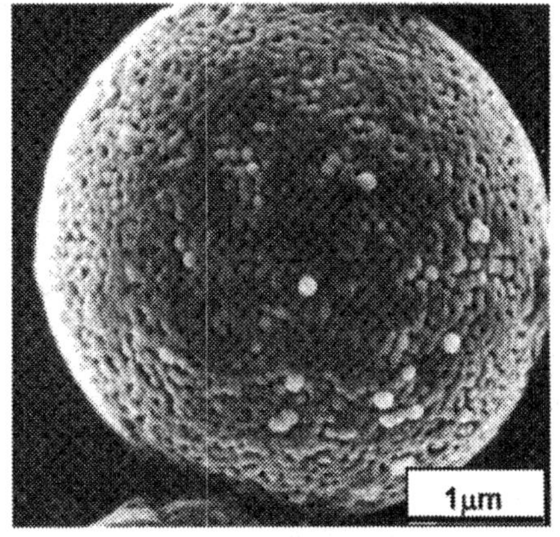

Figure 4, Spherical particle coated with resin particles of fairly uniform size (100 nm). Closed-packed structure of resin particles is formed on the surface of the core particle.

Figure 5, Surface morphology of the processed particles: (a) 90 min processed by only mechanical processing, (b) after annealing at 773 K of the processed particles.

Figure 5 shows the surface morphology of the processed particles of Mg core particles (45μm) and fine amorphous boron particles (0.8μm) by the mechanical processing with the device as shown in Figure 1. Figure 5 (a) shows the surface morphology of 90 min processing with the device, and Figure 5 (b) shows after anneal at 773 K of 10 h in a tube furnace under atmospheric pressure of argon. As is clear from both figures, the crystallites of closed-packed hexagonal MgB_2 are visible on the surface of annealed particle, even though normal reaction temperature of MgB_2 is about 150K higher. Furthermore, DC magnetization (Quantum Design MPMS2) also indicated that superconducting phase was noted around 40 K [4]. These results suggest that the particle bonding with mechanical activation promotes lower temperature reaction as well.

Next, it shows the results in the case of the mechanical and electrical processing of nanoparticles. TiO_2 powder with anatase phase (ST-01, 7nm, Ishihara sangyo kaisha Ltd., Japan) were processed with urea powder in mechanical processing with glow discharge mode. The chamber was exhausted by a vacuum pump to an order of 10^{-3} Pa. Ar gas was introduced. The clearance and the rotation speed were set to 3 mm and 1000 rpm, respectively. The applied electric power was about 50 W and the operating time was 20 min. The processed powder mixture was removed from the device for subsequent annealing at 673 K for 3 hr in an electric furnace to decompose the remained urea. The X ray diffraction pattern indicates that the processed powder was also anatase phase, as is not shown here.

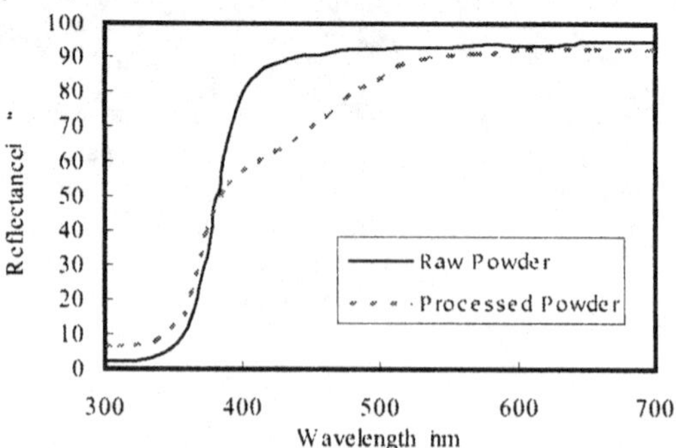

Figure 6, Experimental optical reflectance spectra of the raw and the processed TiO_2 powder.

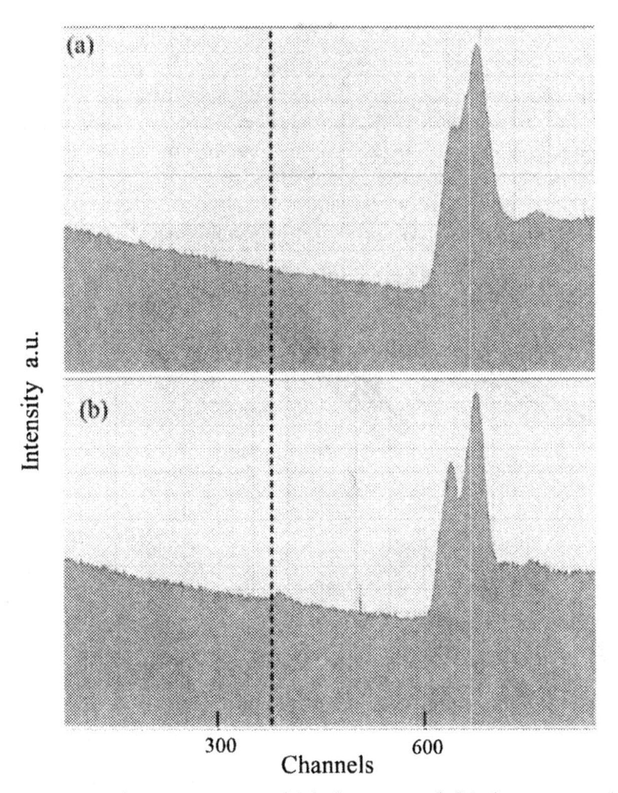

Figure 7, Electron energy loss spectrums of (a) the raw and (b) the processed TiO_2 powder. A dotted line shows the site of Kα edge of nitrogen.

Figure 6 shows the relationships between wavelength and reflectance for the raw and the processed powder. As can be seen, the reflectance of the processed powder drastically decreases the light at less than 500nm, and it indicates the processed powder adsorb the corresponding visible light. Figure 7 shows the electron energy loss spectrums (EELS) of the raw and the processed TiO_2 powder. In the figure, Kα edge of nitrogen is pointed out with a dotted line. Obviously, the Kα edge of nitrogen is observed on the EELS spectrum for the processed powder. These results show that the mechanical processing of TiO_2 and urea powders with Ar glow discharge produces nitrogen-doped TiO_2 powder which may be applicable visible-light photocatalysis [5].

II. NANO/ MICRO STRUCTURE DESIGN OF MATERIALS USING COMPOITE PARTICLE

By making use of composite particles, nano/micro structure was fabricated as shown in Figure 8. This processing route can produce cost-effectively, well-dispersed composite materials with nano or micro sized porous structures. Starting with fumed silica powder, nanosized porous materials can be successfully produced [6]. Commercially available fumed silica has a dendrite structure consisting of primary particles of about 10nm, resulting in several tens of nanometers of pores. By mechanical processing of the fumed silica with glass fibers using the device shown in Figure 1, glass fibers are coated with fumed silica powder. Then, large bulk materials with porous structure of the several tens of nanometers can be easily obtained by subsequent die pressing of the composites. Figure 9 shows micrographs of fumed silica, glass fiber, and fumed silica/glass fiber composite. As can be seen, about 2μm thick fumed silica layer is formed. The fiber reinforced porous materials with 85 % porosity have successfully been fabricated using the composite.

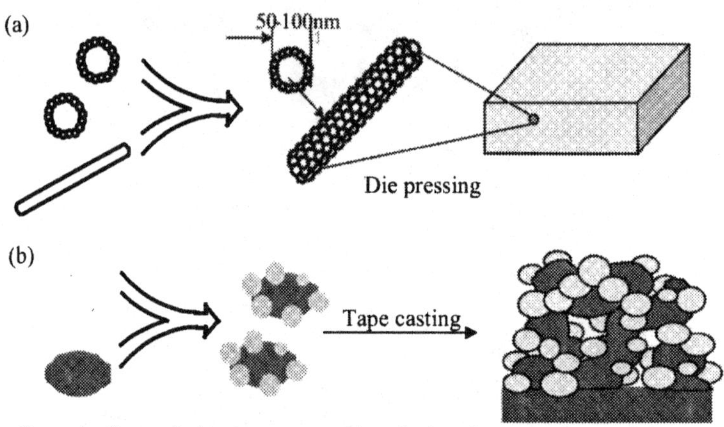

Figure 8, Control of nano/micro structure using designed composite particles (a) for porous material with several tens of nanometers, (b) for micro-scale porous structure, aiming electrode of solid oxide fuel cell.

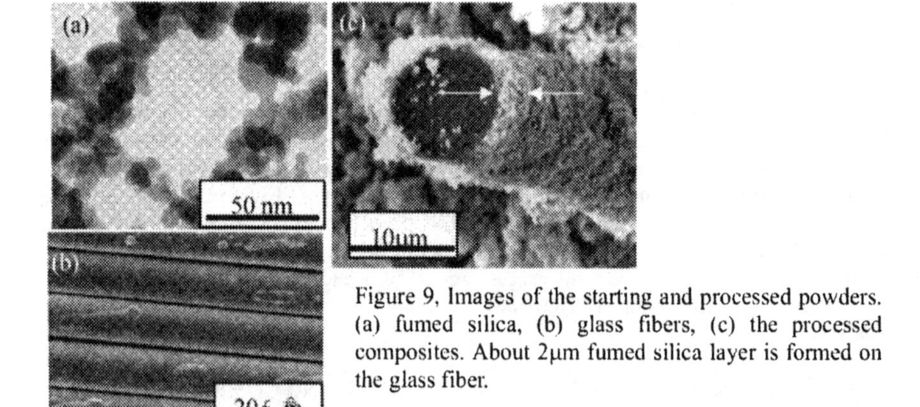

Figure 9, Images of the starting and processed powders. (a) fumed silica, (b) glass fibers, (c) the processed composites. About 2μm fumed silica layer is formed on the glass fiber.

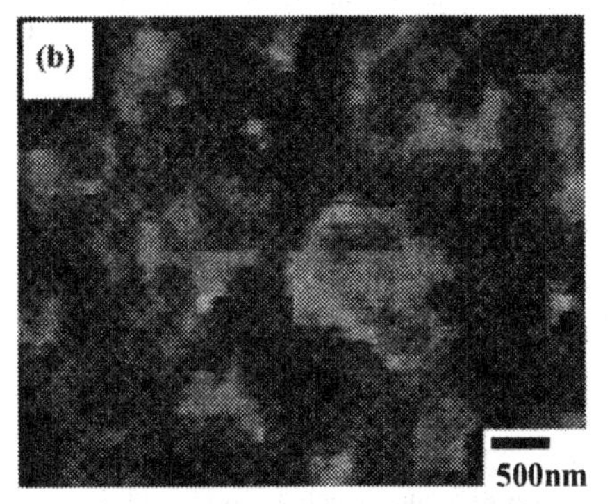

As shown in Figure 10 (b), the well-dispersed composite porous with micro-scale range can be also easily fabricated by selecting the combination of starting powder materials. So far, we have applied this approach to improve anode polarization property of Solid Oxide Fuel Cells (SOFC). Figure 10 shows an example where the microstructure of Ni/YSZ anode electrode of Solid Oxide Fuel Cells (SOFC) is fabricated. In this case, starting composite particles were NiO particles (0.8μm), and YSZ nanoparticles (80nm). The processing was conducted only with mechanical method. Composite particles as shown in Figure 9 (b) were fabricated. Then, they were printed onto the surface of electrolyte, and evaluated the properties of electrode.

Figure 10 (a) shows the microstructure of the anode, (b) shows the mapping image of oxygen by Auger Electron Spectroscopy (AES), and (c) shows that of Ni by AES, respectively. These results make clear that the network structure of Ni and YSZ are well organized to increase the three phase boundary (TPB) length in the electrode. As a result, the anode showed better anode polarization than that of existing electrode made by using ball-milling at 1073 K operation.

Figure 10. Auger Electron Spectroscopy results of a Ni-YSZ cermet anode fabricated from NiO-YSZ composite powder processed by the mechanical method. (a) SEM image, (b) Ni element mapping (c) O element mapping, (*the mapped element is white*)

CONCLUSION

Based on the concept of direct particle bonding technology, a novel mechanical processing apparatus with glow discharge was developed. The apparatus contributed to creating surface design of particles as well as the control of nano/micro structure for better materials.

ACKNOWLEDGEMENT

This work is the result of the Grant-in-Aid for Scientific Research (B) from Ministry of Education, Sports, Culture, Science and Technology of Japan. The authors also wish to acknowledge Prof. Kaneko and Dr. Bun of Kyushu University and Dr. Fukui of Hosokawa Powder Technology Research Institute for helpful assistance and fruitful discussion.

REFERENCE

[1]K.W.Liu, J.S.Zhang, J.G.wang, G.L.Chen, "Formation of nanocomposites in the Ti-Al-Si system by mechanical alloying and subsequent heat treatment", *Journal of Materials Research*, **13**, 1198-1203 (1998)

[2]A.Gulino, R.G.Egdell, I.Fragala, "Mechanically Induced Phase Transformation and Surface Segregation in Bismuth-Doped Tetragonal Zirconia", *Journal of American Ceramic Society*, **81**, 757-759 (1998)

[3]M.Naito, A.Kondo, T.Yokoyama, "Application of Comminution Techniques for the Surface Modification of Powder Materials", *ISIJ International*, **33**, 915-924 (1995)

[4]H.Abe, M.Naito, K.Nogi, M.Matsuda, M.Miyake, S.Ohara, A.Kondo, T.Fukui, "Low Temperature Formation of Superconducting MgB_2 Phase from Elements by Mechanical Milling", *Physica C*, **391**, 211-216 (2003)

[5]R.Asahi, T.Morikawa, T.Ohwaki, K.Aoki, Y.Taga, "Visible-Light Photocatalysis in Nitrogen-Doped Titanium Oxides", *Science*, **293**, 269-271 (2003)

[6]D.Tahara, Y.Itoh, T.Omura, H.Abe, M.Naito, "Formation of Nano Structure on Glass Fiber by Advanced Mechanical Method", *Ceramic Transaction*, (2004) in press.

EVALUATION OF COMPACTION BEHAVIOR BY OBSERVATION OF INTERNAL STRUCTURE IN GRANULES COMPACT

Satoshi Tanaka, Zenji Kato,
Tomohisa Ishikawa, Chia Pin Chiu
Nozomu Uchida and Keizo Uematsu
Department of Chemistry,
Nagaoka University of Technology
1603-1 Kamitomioka, Nagaoka,
940-2188, Japan

Tsutomu Kurita
Technical Management Division,
Japan Nuclear Cycle Development Institute
4-49 Muramatsu, Tokai-mura, Naka-gun,
Ibaraki 319-1184, Japan

ABSTRACT

Compaction curves in each stage were studied by direct observation of internal structure of the compact and the characteristics of spray-dried granules. Three kinds of alumina granules were fabricated by spray drying with changing slurry conditions. The granules were characterized on the shape, the compression strength and the relative density. The compaction curves in uni-axial pressing were measured by using a compression testing machine from 0 to 100 MPa. The internal structures of granules compacts pressed at various pressures were observed using a laser scanning fluorescent microscope. Observation and characteristics of granules indicated that granules started deformation at a yield stress in compaction curves and those granules fractured at the stress of granule strength. The relative density of granules compact did not correspond with the relative density of granules. The result suggests that the rearrangement on particles in the granule start before the finishing granule deformation.

INTRODUCTION

Fundamental understanding of die compaction processing is very important in ceramics, since the process affects the structure and properties of ceramics significantly[1,2]. Especially, the mechanism for the transformation from the granules to compact needs to be fully understood. Our understanding on the mechanism is based mostly on the analysis of the compaction curve (Fig.1), namely the relation between the density of the pressed body and the stress applied on the punch[3-20]. The compaction curve consists of three stages which are accompanied by the changes of structure; the rearrangement of granules with the density of granules unchanged in stage 1, deformation of granules with their density kept constant to fill spaces between them in stage 2, and rearrangement of primary particle within granules in stage 3. However, there are little direct understandings on the change of structure at the transition of these stages and the relevance between the strength of granule and

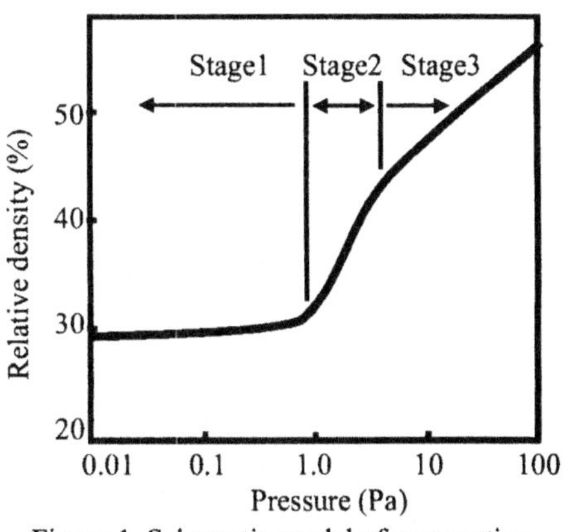

Figure 1. Schematic model of compaction curve

the stress for transition[22,23]. The aims of this study are to examine the structure change in the compaction process with a new characterization technique and to relate the transition stress from stage 1 to stage 2 to the strength of granules.

EXPERIMENTAL PROCEDURE

(1) Preparation of granules

A commercial alumina powder (AL160-SG1 Showa Denko K.K., Japan) was used in this study. The reported median particle size was 0.6μm, which was measured with the x-ray sedimentation method[24]. The specific surface area was 6.0m^2/g. The powder was mixed with distilled water and dispersant (ammonium polyacrillic acid: Seruna D305, Chukyoyushi Japan) in a ball mill for 20h. The concentration of the dispersant was adjusted to prepare dispersed and flocculated slurries. The powder contents in slurry were 20 and 30vol%. The slurries were treated in a vacuum and dried in a spray-dryer□Model SD13, Mitsui-Mining Co., Japan□to prepare the granules. The operating conditions of the dryer were: the inlet air temperature 200 °C, the atomizer speed 10000rpm and the slurry feed rate 30g/min. The granules were sieved to adjust their size 26μm to 38μm.

(2) Characterization of granules

The granules were kept in an atmosphere of controlled relative humidity 70% for at least 1 day before use. The compression strength of a granule was measured by a micro compression testing machine (MCTM500, Shimadzu, Japan) with flat indenter(ϕ=50μm). The compression strength St was calculated using the following equation[25-30],

$$St = 1.4\frac{F}{2\pi d^2} \tag{1}$$

where, F is the fracture force, and d the diameter of a granule. The mean compression strength of a granule was calculated from ten granules in this study. The relative density of a granule was calculated from the porosity of a granule, which was measured by the mercury intrusion porosimetry (Poresizer9310, Micromeritics, USA)[3]. The shapes of granules were also examined by the liquid immersion method[1].

(3) Compaction curve and observation of granules compact

The spray-dried granules were uniaxially pressed in double action by a universal mechanical testing machine (Autograph AG1 Shimadzu Japan) at crosshead speed 0.5mm/min. The compaction curves were constructed for each granule from the load and displacement recorded automatically. The relative density (R.D.) of the granules compact during compression test was measured using a following

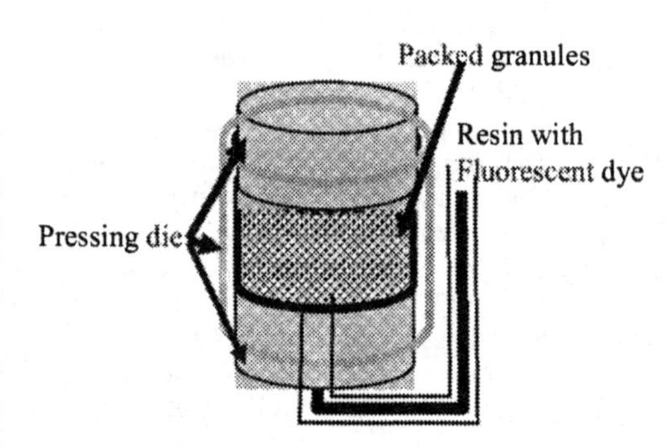

Figure 2. Solidification of pressed granules

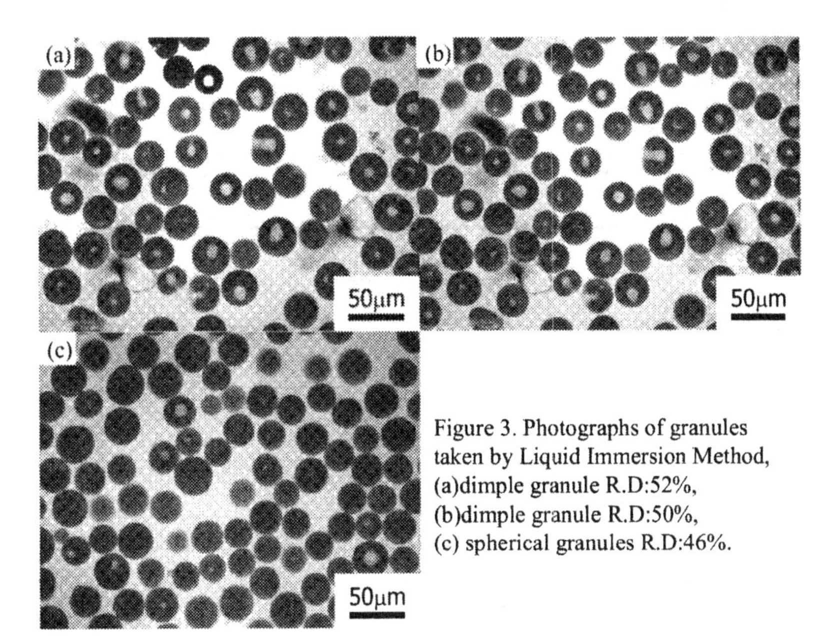

Figure 3. Photographs of granules taken by Liquid Immersion Method, (a)dimple granule R.D:52%, (b)dimple granule R.D:50%, (c) spherical granules R.D:46%.

equation,

$$R.D. = 100\frac{w/Ah}{\rho} \qquad (2)$$

where, w was the weight of filled granules, A the base area of the compact (cross section area of pressing punch), h the height of the compact in the die, and ρ the theoretical density of alumina: 3.987×10^3 kg/m^3. The resin (PETROPOXY 154 RESIN, Palouse Petro Products, USA) containing fluorescent dye was impregnated into the compact as shown in Fig.2 and was cured at 80°C for solidification. The shrinkage in solidification was less than 1%. The cross sections of solidified compact were polished with diamond slurry and were observed with a laser comfocal scanning fluorescent microscope. In this mode of observation, the spaces between particles and granules were selectively observed [24,31].

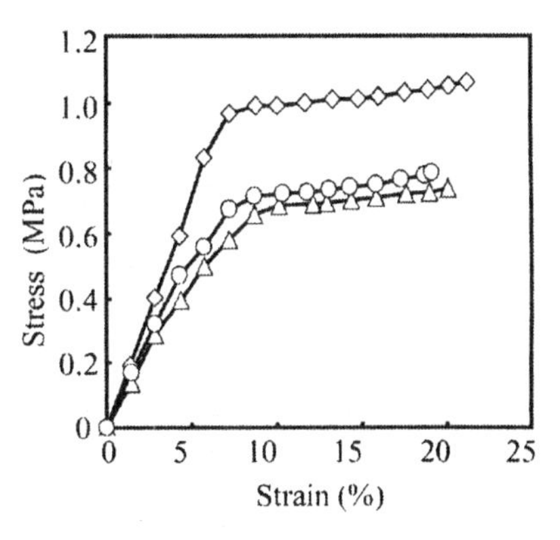

Figure 4. Stress-strain curves of a granules,
◇:dimple granule R.D.52%,
△: dimple granule R.D.50%,
○: spherical granules R.D.46%.

Table 1. Characteristics of granules

Slurry	Shape	Relative Density	Compression strength
30vol%, dispersed	Dimple	52%	1.2±0.3 MPa
20vol%, dispersed	Dimple	50%	0.7±0.1 MPa
20vol%, non-dispersed	Spherical	46%	0.6±0.1 MPa

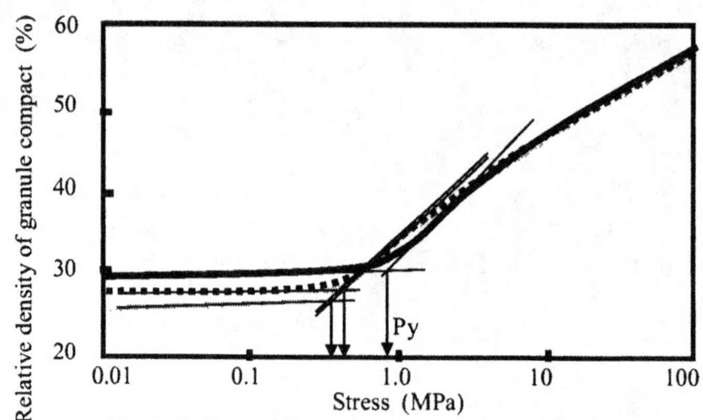

Figure 5. Compaction curves of pressing granules,
Solid line : dimple granule, RD:52%,
Broken line: dimple granule RD:50%
Gray solid line: spherical granules, RD: 46%.

RESULTS

Fig.3 shows the photomicrographs of granules prepared in this study. The granules prepared from well-dispersed slurry have a dimple (Fig. 3(a) and (b)), whereas those prepared from the flocculated slurry have a spherical shape (Fig. 3 (c)).

Table 1 shows the characteristics of granules prepared in this study. The relative density (RD) appeared to increase with increasing powder content in the slurry; they were 52% and 50% from slurries of 30vol% and 20vol%, respectively. The relative density was higher for the granules with dimple than those of spherical shape. The relative density of spherical granules was 46 %.

Fig.4 shows typical stress-strain curves in granule compression tests. The strength was represented by the yield stresses with equation (1). The granules with a dimple (RD:52%) have the highest strength in all granules. The strength of other granules is almost the same.

Fig.5 shows the compaction curves in die pressing for these granules. The compaction curves vary with the types of granules. The initial packing density increased with increasing density of granules. The relative density of the compact body increased sharply at the yield stress P_y about 1 MPa. The slope of the curve changed again between 1 to 10 MPa.

Fig.6 shows the confocal laser scanning fluorescent micrograph of the compact made with the granules prepared from the dispersed slurry of solid content 30 vol%. The granules were packed loosely in the pressing die before pressing. Their shapes are the same to those of granules. With applied stress, the granules contact each other at pressure around 1 MPa. Then, some granules

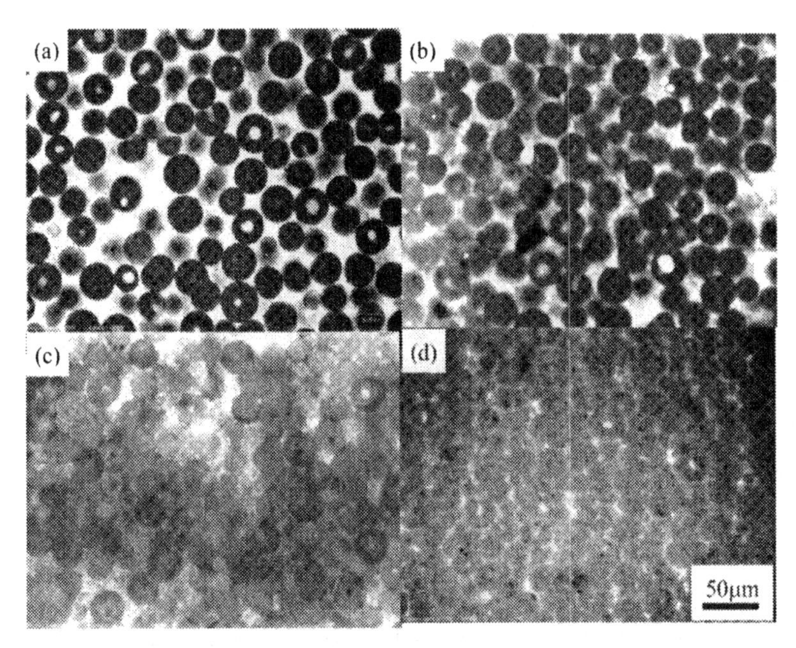

Figure 6. Internal structure of the green body during compaction taken by laser scanning fluorescent microscopy, (a) packing, (b) after pressing at 1MPa, (c) after pressing at 2 MPa. (b) after pressing at 100 MPa,

started to deform at 2 MPa. The granules were highly deformed and densely packed in the compact made at 100MPa.

DISCUSSION

The compaction curves change with the characteristics of granules. In the first stage, granules are packed loosely in the pressing die as shown in Fig.6. A significant change of structure suggests the rearrangement of granule in the stress region below P_y. In the second stage, the granules were reported to deform and fracture in many researches. However, it has not been understood well when granules fractured in this stage. The granule fracture is similar to the yield stress P_y in compaction curve.

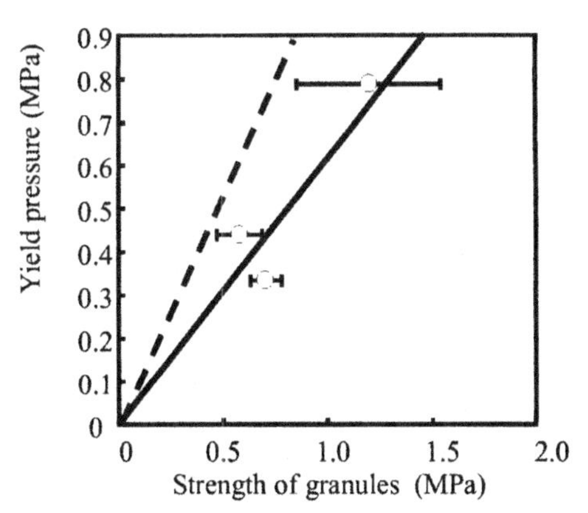

Figure 7. The relationship between granule strength and yield pressure in the compaction curves.

Fig.7 shows the relationship between the granule strength S_t and the yield stress P_y in compaction curve. The yield stress increased with increasing granule strength. This result is easily understood. However, the granules strength S_t is always larger than the yield strength P_y. The granules start to deform at the yield stress and then fracture when applied stress reaches the granule strength. In Fig.6(b), granules only contacted each other and deformed slightly at the applied stress 1 MPa, where the stress was larger than the yield stress.

Fig.8 shows the relationship between the relative density of granules and granule compact. It was believed that the deformation of granules was complete at the end of the second stage and the primary particles in the granules started to rearrange in the compact in the third stage. However, the relative density of the compact at the changing point from stage 2 to stage 3 does not correspond to the relative density of the granules. The relative density of the compact is smaller than the relative density of granules as indicated in Fig.8. The result suggests that the primary particle starts rearrangement in the granules before granule deformation is finished. The detailed analysis of rearrangement in the granules will be reported in a subsequent study.

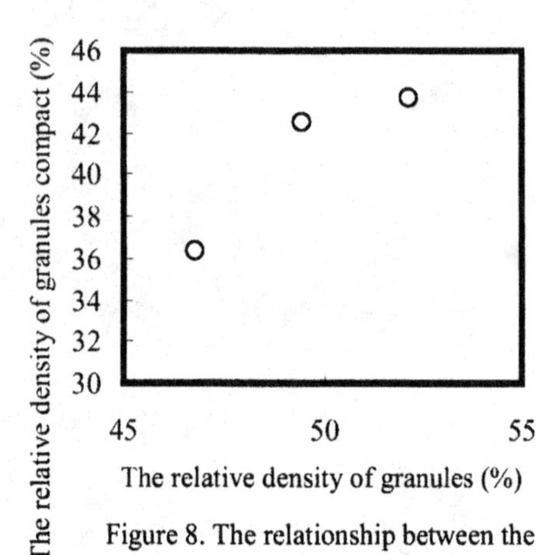

Figure 8. The relationship between the relative density of granule compact and granule.

CONCLUSIONS

Compaction curves in each stage were studied by direct observation of internal structure of the compact and the characteristics of spray-dried granules. Three kinds of alumina granules were fabricated by spray drying with changing slurry conditions. The granules were characterized on the shape, the compression strength and the relative density. The compaction curves in uni-axial pressing were measured by using a compression testing machine from 0 to 100 MPa. The internal structures of granules compacts pressed at various pressures were observed using a laser scanning fluorescent microscope. The following conclusions were obtained.

(1) Observation and characteristics of granules indicated that granules started deformation at a yield stress in compaction curves. Granules fractured at the stress of granule strength.

(2) The relative density of granules compact did not correspond with the relative density of granules. The result suggests that the rearrangement of particles in the granule start before the final granule deformation.

ACKNOWLEDGMENTS

This research was partly supported by COE program Grant-in-Aid for Scientific Research from the Japan Ministry of Education, Culture, Sports, Science and Technology.

REFERENCE

[1]K.Uematsu, M.Miyashita, J.Y.Kim, Z.Kato, and N.Uchida, "Effect of Forming Pressure on the Internal Structure of Alumina Green Bodies Examined with Immersion Liquid Technique", *Journal of the American Ceramic Society*, 74 2170-74 (1991).

[2]W.J.Walker,Jr., J.S.Reed and S.K.Verma, "Influence of Granule Character on Strength and Weibull Modulus of Sintered Alumina", *Journal of the American Ceramic Society*, 82 50-56 (1999).

[3]W.J.Walker,Jr., J.S.Reed and S.K.Verma, "Influence of Slurry Parameters on the Characteristics of Spray-Dried Granules", *Journal of the American Ceramic Society*, 82 1711-19 (1999).

[4]A.R.Cooper.Jr. and L.E.Eaton, "Compaction Behavior of Several Ceramic Powders," *Journal of the American Ceramic Society*, 45 97-101 (1962).

[5]D.E.Niesz, R.B.Bennett and M.J.Snyder, "Strength Characterization of Powder Aggregates", *American Ceramic Society Bulletin*, 51 677-80 (1972).

[6]S.J.Lukasiewicz and J.S.Reed, "Character and Compaction Response of Spray-Dried Agglomerates", *American Ceramic Society Bulletin*, 57 769-801 (1978).

[7]R.A.Thompson, "Mechanics of Powder Pressing: I, Model for powder dinsification", *American Ceramic Society Bulletin*, 60 237-43 (1981).

[8]R.A.Dimilia and J.S.Reed, "Stress Transmission during the Compaction of a Spray-Dried Alumina Powder in a Steel Die", *Journal of the American Ceramic Society*, 66 667-72 (1983).

[9]R.A.Dimilia and J.S.Reed, "Dependence of Compaction on the Tg of the Binder Phase" *Journal of the American Ceramic Society*, 62 484-88 (1983).

[10]R.G.Frey and J.W.Halloran, "Compaction Behavior of Spray-Dried Alumina", *Journal of the American Ceramic Society*, 67 199-203 (1984).

[11]U.Klemm and D.Sobek, "Influence of Admixing of Lubricants on Compressibility and Compatibility of Uranium Dioxide Powders", *Powder Technology*, 57 135-42 (1989).

[12]J.S.Reed, "Principles of Ceramics Processing 2nd edit." John Wiley Inc., 418-49 (1995).

[13]B.D.Mosserm J.S.Reed and J.R.Varner, "Strength and Weibull Modulus of Sintered Compacts of Spray Dried Granules," *Journal of the American Ceramic Society*, 71 105-9 (1992).

[14]J.H.Song and R.G..Evans, "A Die Pressing Test for the Estimation of Agglomerate Strength", *Journal of the American Ceramic Society*, 77 806-814 (1994).

[15]P.R.Mort R.Sabia, D.E.Niesz and R.E.Riman, "Automated Generation and Analysis of Powder Compaction Diagram", *Powder Technology*, 79 111-19 (1994).

[16]D.W.Whitman, D,.I.Cumbers and X.K.Wu, "Humidity Sensitivity of Dry-Press Binders" *American Ceramic Society Bulletin*, 74 76-79 (1995).

[17]R.D.Carneim and G.L.Messing, "Response of Granular Powders to Uniaxial Loading and Unloading", *Powder Technology*, 115 131-38 (2001).

[18]L.J.Neergaard and M.B.Nawaz, "Dry-Pressing Behavior of Silicon-Coated Alumina Powders", *Powder Technology*, 98 104-8 (1998).

[19]N.Shinohara, S.Katori, M.Okumiya, T.Hotta, K.Nakahira, M.Naito, Y.I.Cho and K.Uematsu, "Effect of Heat Treatment of Alumina Granules on the Compaction Behavior and Properties of

Green and Sintered Bodies", *Journal of the European Ceramics Society*, 22 2841-48 (2002).
[20]S.J.Glass, K.G.Ewsuk and M.J.Readey, "Ceramic Granule Strength Variability and Compaction Behavior", *27th International SAMPE Technology Conference*, 635-44 (1995).
[21]F.M.Mahoney and M.J.Readey, "Applied Mechanics Modeling of Granulated Ceramic Powder Compaction", *27th International SAMPE Technology Conference*, 645-57(1995).
[22]T.A.Deis and J.J.Lannutti, "X-ray Computed Tomography for Evaluation of Density Gradient Formation during the Compaction of Spray-Dried Granules", *Journal of the American Ceramics Society*, 81 1237-47 (1998).
[23]C.M.Kong and J.J.Lannutti, "Localized Densification during the Compaction of Alumina Granules: The Stage I-II Transition", *Journal of the American Ceramics Society*, 83 685-90 (2000).
[24]S.Tanaka, Z.Kato, N.Uchida and K.Ueamtsu, "Direct Observation of Aggregates and Agglomerates in Alumina Granules, *Powder Technology*, 129 153-55 (2003).
[25]Y.Hiramatsu, Y.Oka and H.Kiyama, "Rapid Determination of the Tensile Strength of Rocks with Irregurar Test Pieces", *Nihon kogyokai-shi*, 81 1024-30 (1965).
[26]H.Takahashi, N.Shinohara, K.Uematsu and J.Tsubaki, "Influence of Granule Character and Compaction on the Mechanical Properties of Sintered Silicon Nitride", *Journal of the American Ceramics Society*, 79 843-48 (1996).
[27]H.Kamiya, M.Naito, T.Hotta, K.Isomura, J.Tsubaki and K.Uematsu, "Powder Processing for Fabricating Si_3N_4 Ceramics", *American Ceramics Society Bulletin*, 76 [10] 79-82 (1997).
[28]Y.Yamamoto, "Micro Compression Test for Ceramics Powder", *Bulletin of the Ceramic Society of Japan*, 37 74-76 (2002).
[29]J.S.Reed, "Principles of ceramics processing 2nd edit." John Wiley Inc., 247-76 (1995).
[30]N.Miyata, Y.Ishida, T.Shiokay and Y.Matsuo, "The effect of Characteristics of Compressive Deformation of Ceramic Granules on CIP Compaction Behavior and Sinterability (Part I)", *Journal of the Ceramics Society of Japan*, 1275-81 (1995).
[31]Y.Saito, S.Tanaka, N.Uchida and K.Uematsu, "Direct Evidence for Low-Density Regions in Compacted Spray-Dried Powders", *Journal of the American Ceramics Society*, 2454-56 (2001).

Microstructure Characterization and Modeling

GEOMETRY OF MICROSTRUCTURAL EVOLUTION IN SIMPLE SINTERING

Robert T. DeHoff
Department of Materials Science and Engineering
University of Florida
Gainesville, FL 32611

ABSTRACT

The evolution of microstructure that is sintering is observed in the interactive changes in the geometric properties (volume, surface area and curvature, line length) of five microstructural features: the porosity (p), the pore-solid interface (αp), the grain boundaries (αα), the triple lines where the grain boundaries intersect the pore-solid interface (ααp), and triple lines in the grain boundary network, (ααα). These property changes result from processes (densification, surface rounding, grain boundary migration), that are mutually influential. Each of these processes operates under its own thermodynamic driving force, which in simple sintering is determined by the curvatures of the surfaces involved. This driving force activates mechanisms that produce the migration of defects between sources and sinks and along paths that are appropriate to that process. Changes in geometric properties that result from each process influence the driving forces, sources, sinks and paths of the others. Surface rounding decreases curvature and hence the driving force for densification. Densification at some stage produces a grain boundary network that is sufficiently connected for grain boundary migration to begin. The coarsening grain boundary network later determines the length scales of the paths that control densification. Models for sintering that do not recognize these interactions are at best incomplete, and at worst misleading.

INTRODUCTION

Simple sintering, which in this context means sintering of a single phase crystalline material under the driving force of surface tension unaided by applied mechanical forces, is complicated enough. From a purely geometric point of view a stack of powder consists of two interpenetrating multiply connected networks: the solid phase and the porosity. Upon heating to a significant fraction of its melting point this initially complex geometry changes, nucleating and growing a grain boundary network which plays a central role in the evolution of the topology of the pore and solid structures, and particularly in the accompanying densification of the mass. This paper reviews the geometry of this evolution and reviews a model based upon two related tessellations (space filling cell structures) of the microstructure [1, 2] that permits the visualization of the overall process from beginning to end.

FEATURE SETS IN THE MICROSTRUCTURE

Figure 1 is a sketch of a micrograph of a partially sintered single phase material. Four geometric feature sets are visible on this section through the three dimensional microstructure:

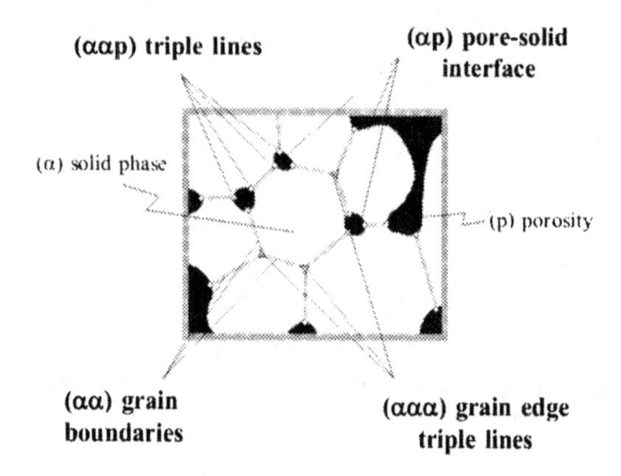

Figure 1. Sketch of a cross section through a partially sintered structure showing intersections with the solid phase (α), the pore phase (p), pore - solid interface (αp), grain boundaries (αα), ααp triple lines, and ααα grain edges.

- α - the solid phase (three dimensional feature set shown in white);
- p - the pore phase (three dimensional feature set shown in black);
- αp - pore-solid interface (two dimensional set of surfaces bounding the pore and solid phases);
- αα - grain boundary (two dimensional set of surfaces separating grains in the solid phase);
- ααα - grain edges (one dimensional set of triple lines where three grains meet);
- ααp - solid-solid-pore edges (one dimensional triple line where grain boundaries meet the pore-solid interface)

Measures of the quantities of each of these geometric feature sets (volume, surface area, line length) may be obtained by applying the standard methods of stereology. Measures of the connectivity of the porosity and curvature related properties of the pore-solid interface are also available. In principle, the path of microstructural change (sequence of geometric states attained during the process), the kinematics of this geometrical evolution and the kinetics of the process may be quantified by monitoring these properties

MICROSTRUCTURAL EVOLUTION

Initially the geometry of the pore-solid interface at points of contact between pairs of particles is highly unstable; necks form and grow at these contact points. The feature sets that exist initially are α, p, and αp, Figure 2a. Since the mating particles are crystalline a new interface, - a grain boundary - , forms and grow at each neck, Figure 2b.. New feature sets are thus introduced: αα and ααp, the circular line bounding each neck, Figure 2c. These changes are generally accompanied by densification, a macroscopic decrease in the volume of the system. At the microscopic level densification manifests itself as a decrease in the distance between adjacent

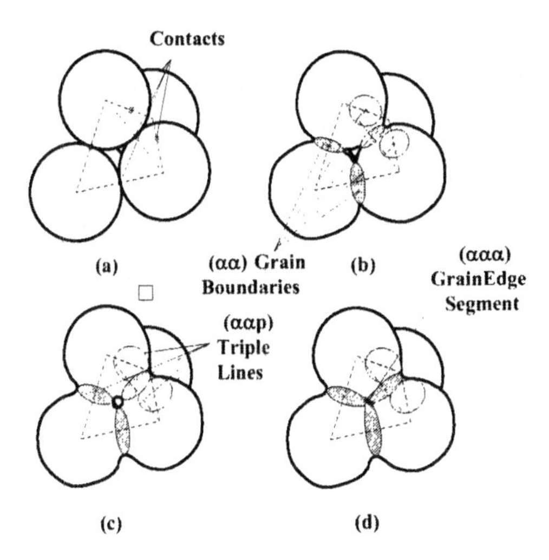

Figure 2. Particles in a powder stack touch at contacts (a) which grow to form necks which contain grain boundaries (b). These grain boundaries intersect the pore solid (αp) surface at (ααp) triple lines (c). Continued neck growth pinches off channels in the pore network, simultaneously forming (ααα) grain edges (d).

particle centroids. The structure more or less uniformly pulls itself together. Nonuniformities in this process are of central interest in modeling sintering because these regions may be at the heart of performance problems in the finished product.

As the process continues necks continue to grow: Adjacent necks on the same particle impinge as their ααp triple lines touch, producing complexities in the local geometry of the associated grain boundary areas. Sets of three particles in mutual contact surround an elongated channel in the pore network, Figure 2b. As the three necks between these particle pairs impinge, the channel pinches off (the channel attains a critical shape that renders it unstable from a capillarity point of view), Figure 2c. The meeting of the three necks produces an element of a new geometric feature set, a grain edge (an ααα triple line). At each end of this element of edge is an αααp quadruple point where the grain edge intersects the pore surface, Figure 2d.

Each such channel closure event reduces the connectivity of the pore network by one unit. At later stages channels formed between four (and more) particles achieve unstable shapes and close. The pore network gradually becomes less connected, and, eventually, disconnected. Densification proceeds apace. In the process the grain boundary network, which is initially completely disconnected, becomes progressively more connected.

GRAIN CELL STRUCTURE VISUALIZATION OF THE OVERALL PROCESS

This process may be visualized on the basis of the cell model shown in Figure 3. Initially a cell is defined (perhaps somewhat ambiguously) as containing a single particle (grain) with its "associated porosity", Figure 3a. At time zero the cell faces are empty, except for some points of contact between particles. As necks grow, the cell faces become progressively more occupied with grain boundary area, Figure 3b, c, d. Formation of grain edges ($\alpha\alpha\alpha$) (and the associated channel closure process) is seen to occur when the expanding grain boundary areas contact the edges in the cell model structure. As more edges are contacted the grain boundary structure on the cell faces transitions from a collection of disconnected areas each surrounded by porosity, Figure 3b, toward a structure with isolated areas of porosity in a connected grain boundary network, Figure 3e. At full density the cell structure is completely covered by grain boundary.

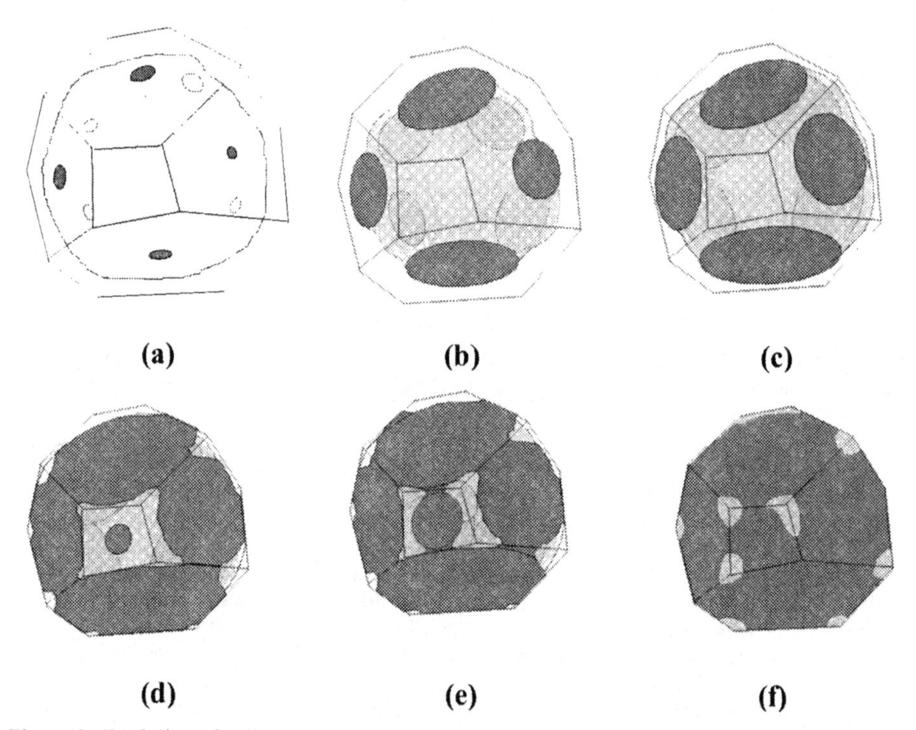

Figure 3. Evolution of cell (a particle and its " associated porosity" during sintering. Necks form with grain boundaries at interparticle contacts (a) and grow (b), (c) until adjacent necks impinge at cell edges to form $\alpha\alpha\alpha$ triple lines in the grain structure at cell edges (d). Cell faces become progressively more completely covered (e) eventually isolating pores at the cell corners (f). With further densification these isolated pores may disappear.

This evolution is complicated by the fact that at some intermediate stage the grain boundary network becomes sufficiently connected so that it may begin to move, tearing away from the pinning porosity, and producing a spontaneous increase in the scale of the cell structure in Figure 3. Grain growth and densification are geometrically coupled but mechanistically independent. Grain boundaries move in response to the local curvature of the αα surface and involve subtle motions of atoms near the grain boundary. This motion may be constrained by interactions involving the αp interface. Densification operates in response to the local curvature of the αp interface and involves diffusion (along the grain boundary or through the volume of the adjacent crystal) over length scales that are at least partially determined by the structure of the αα interfaces in the system. The timing of the grain growth process with respect to densification may have a significant effect on the microstructural state ultimately attained by the system.

BIPYRAMID CELL STRUCTURE AND DENSIFICATION

The grain cell structure described above provides a basis for the continuous visualization of the evolution of the pore and grain boundary structures from the beginning to the end of the process. The further visualization of the flows necessary to produce densification may be aided by the construction of a second cell structure within the grain cell structure. This bipyramid cell structure is constructed beginning with the faces in the grain cell structure. Figure 4 shows a polygonal face in the grain cell structure. Each face is uniquely associated with a particle pair. Locate the centroids of the two particles and construct straight lines from each centroid to the corners of the face. The volume enclosed by this construction is a pyramid with the grain cell face as a base and the perpendicular distance from the centroid to the face as the altitude. The pair of pyramids constructed in this pair of particles is a bipyramid cell. The collection of bipyramid cells is also space filling, being a further subdivision of the grain cell structure.

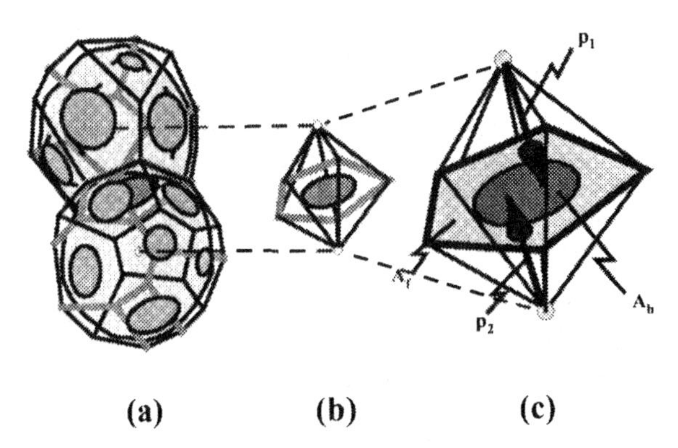

Figure 4. Every cell face has an associated bipyramid construction (a): two pyramids joined at the face (b) with area A_f partially covered by grain boundary with area A_b, (c). The altitudes p_1 and p_2 decrease as shrinkage occurs as a result of the annihilation of vacancies at the grain boundary

Locally, densification is the motion of paired particle centers (apexes of the bipyramid) toward each other.

The base of the bipyramid initially contains the point of contact between the particle pair. (A face may be initially empty because the particle pair associated with it are not in contact.) As time goes on and neck growth occurs the face is increasingly covered by $\alpha\alpha$ grain boundary.

The thermodynamics of curved surfaces shows that elements of αp surface that have negative mean curvature develop an increase in local vacancy concentration over that associated with a flat grain boundary. This sets up a difference in vacancy concentration in the solid phase which drives the flow of vacancies from near the αp surface to the grain boundary, where they can be annihilated, Figure 5. There is a competition for control of this flow by diffusion through the volume of the crystal and diffusion along the grain boundary.

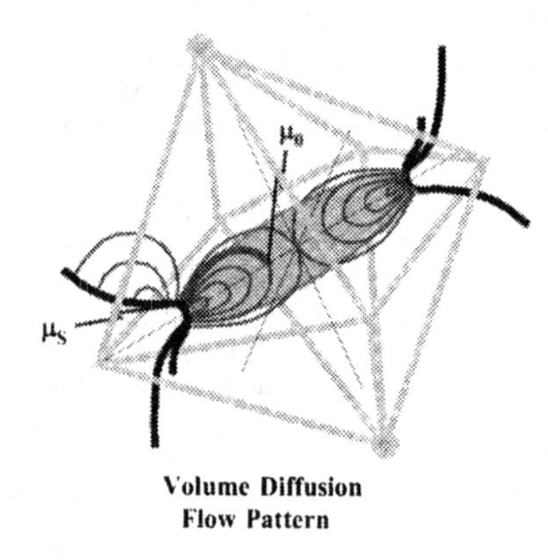

Figure 5. Flow geometry in a bipyramid. The negative mean curvature of elements of the αp surface near $\alpha\alpha p$ triple lines yield in increase in the chemical potential of vacancies μ_s compared the value at the grain boundary, μ_0. This chemical potential difference drives the flow of vacancies from the αp surface to points distributed on the grain boundary, where their annihilation produces densification.

The annihilation of a layer of vacancies at the grain boundary on the bipyramid base is effectively the removal of a layer of atoms between the particle centers. The vacancy annihilation process thus controls center-to-center shrinkage in each bipyramid locally, and hence densification globally. It is expected that this annihilation process proceeds continuously. It might be useful to think of a continuous painting of the grain boundaries in the structure with vacancies for annihilation. Over time the annihilation events must be uniformly distributed over the grain boundary area in each bipyramid. This principle operates over the full range of the process. Densification results by painting vacancy annihilation events uniformly over the area of the grain faces in the system.

This geometry dictates the diffusion length scales involved in the process, a central factor in the kinetics of the process. The vacancy source remains distributed at the curved αp interface. The sinks are uniformly distributed over the associated grain boundary faces. At the beginning of the process the mean curvatures at the neck boundaries are large and negative and the driving force is high. At the same time the diffusion length scales, which are of the order of the radius of the necks, are short. Thus, gradients are high and the flows are large relative to later stages in the process. Densification rates are highest at the outset of the process and slow down as αp curvatures diminish and diffusion length scales increase.

A second factor contributes to further increase early densification rates: the efficiency of the vacancy annihilation process. The annihilation of an effective layer of vacancies at the grain boundary on the face of the ith bipyramid decreases the height of the bipyramid by some amount, dp_i. The volume of the bipyramid is decreased by $A_{fi}\, dp_i$, where A_{fi} is the area of the cell face. The volume associated with the number of the vacancies annihilated in this step is given by $A_{bi}\, dp_i$, where A_{bi} is the area of the grain boundary in the ith face. The vacancy annihilation process is thus leveraged by an efficiency factor, A_{fi} / A_{bi}. At the beginning of the process when the grain boundary occupies a small fraction of the cell face the annihilation of a few vacancies produces a large volume change. As sintering proceeds and the grain boundaries cover progressively more the cell faces this efficiency factors decreases toward a value of one at the end of the process.

NONUNIFORMITIES IN THE GEOMETRY OF THE CELL STRUCTURES

In real powder stacks there are variations in particle shape, particle size and stacking. Individual bipyramid cells cannot evolve independently; they are an element in a contiguous mass of cells. They are mechanically connected to their neighbors in three dimensions. For example the bipyramid associated with a contact between a pair of small particles in a stack with a size distribution would have a high shrinkage rate if it could act independently, Figure 6.

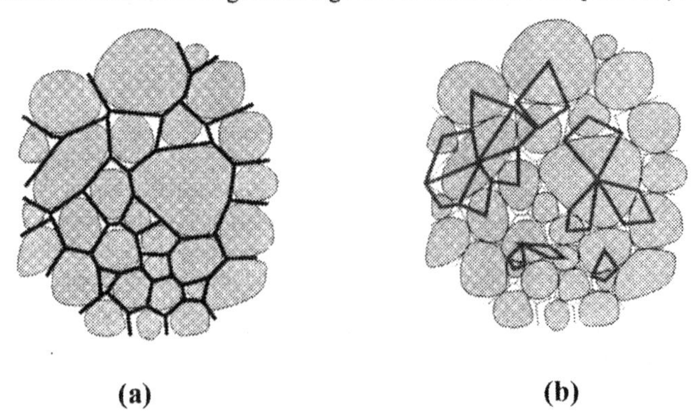

Figure 6. Two dimensional sketch of the cell structure associated with a stack of particles with a size distribution (a). Some bipyramid (bitriangles in two dimensions) cells are illustrated in (b)

Neighboring bipyramids may be larger and therefore shrink more slowly. The shrinkage of the small bipyramid is constrained by that of its neighbors. In the local flow behavior these constraints may act to reduce the compressive stresses at the neck between the fine particles, increasing the chemical potential of vacancies toward the value at the adjacent $\forall p$ interface, and reducing the driving force for diffusion and thus the rate of densification.

Stacking flaws that produce local regions with excess porosity are characterized by bipyramids between particle pairs that are not in contact, Figure 6. These bipyramids shrink only as a result of their connections with adjacent cells that do contain grain boundary on the face and have the capability to annihilated vacancies and produce local shrinkage. For each bipyramid, motion of the particle centroid toward each other removes a volume of porosity that is proportional to the whole face area.

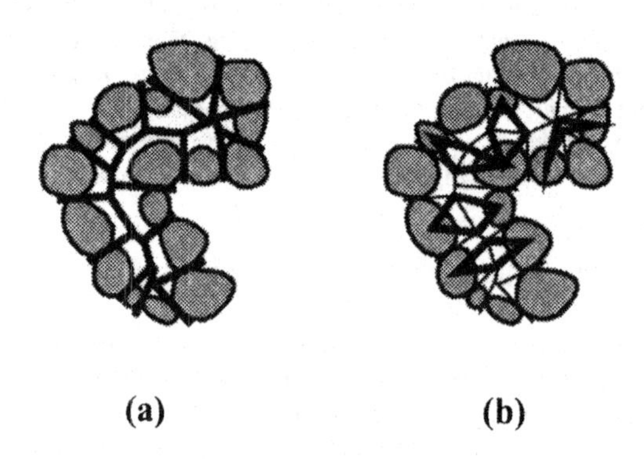

Figure 6. Two dimensional sketch of a large pore resulting from flaws in the powder stack. Bipyramid cells (bitriangles in two dimensions) associated with the flaw do not contain grain boundaries.

In the later stages of the sintering process as the grain boundary network connects to itself grain boundary migration may begin. In order for grains to grow, their number must decrease. This occurs by the annihilation of simple grains (e.g., tetrahedral grains) that exist in the grain structure. The geometry of the topological events that lead to this grain annihilation process is complex, particularly when interactions with the remaining pore structure are considered. Anomalies in particle stacking will also lead to anomalies in the grain growth process. Incorporation of grain growth into the description of the sintering process is a challenge that is yet to be met, but the competition between grain growth and densification is a key factor in determining the ultimate microstructure attained in sintering.

SUMMARY

Two nested cell structures provide a potentially useful visualization tool for the description of simple sintering from beginning to end. Considerations based upon these cell structures provide a basis for visualizing sources, sinks and paths for mass transport, the related driving forces and ultimately the rates of transport and the associated rates of geometric evolution. These cell structures may provide a basis for finite element analysis of this process that incorporates particle and stacking variables. The evolution of the competition between densification and grain growth remains a challenging and important problem.

REFERENCES

1. R.T. DeHoff, "A Cell Model for Microstructural Evolution During Sintering," 23 - 39, in *Sintering and Heterogeneous Catalysis*, Plenum Press, New York 1984.

2. R.T. DeHoff, "A Stereological Model of Sintering," pp 55-72., in *Science of Sintering. New Directions for Materials Processing and Microstructural Control*. (D.P Uskokovic, H.Palmour III and R.M. Spriggs, Eds.) Plenum Press, New York, 1989.

IN SITU OBSERVATION OF SINTERING BEHAVIOR IN BARIUM TITANATE USING AN ENVIRONMENTAL SCANNING ELECTRON MICROSCOPE

Srinivas Subramaniam*, Rodney Roseman,
497 Rhodes Hall, M.L. 0012
Department of Chemical and Materials Engineering,
University of Cincinnati
Cincinnati, OH 45221-0012

ABSTRACT

In situ studies of the high temperature sintering behavior of undoped barium titanate have been carried out. An environmental scanning electron microscope (ESEM) equipped with a 1500°C hot-stage furnace accessory has been used to perform the experiments. Barium titanate powders (Ba/Ti ratio 1.004) processed using conventional ceramic routes were sintered in the ESEM at temperatures between 1320°C and 1375°C. No noticeable changes are observed to occur in the materials at temperatures between 500°C and 1100°C. Between 1100-1200°C, an abrupt change in microstructure can be observed with the formation of two dimensional structures which appear to be needle-like in shape. As the samples are heated beyond this stage, these structures are observed to develop into anisotropic hexagonal disks with their final dimensions dependant on the sintering temperature. Solid-state sintering is seen as the dominant mechanism occurring in these studies. This said, the presence of discontinuous liquid phase regions can be observed in the microstructures of samples which have been sintered at 1350°C and 1375°C. These results are in agreement with studies which attribute abnormal growth mechanisms to the presence of inhomogeneous Ti-excess regions in Ba-excess powders with Ba/Ti ratios $\leq$ 1.006. These studies show the feasibility of using ESEM techniques to study dynamic high temperature processes occurring in ceramics.

INTRODUCTION

Advanced ceramic materials used in structural and electronic applications must meet stringent performance requirements in industry. A fundamental understanding of materials and processes is of paramount importance if consistent performance and reliability of these materials is to be achieved. Barium titanate materials are widely used in research and industry for applications in dielectric and sensor technologies. Though the industrial significance of these materials has resulted in extensive research on their properties and applications, a comprehensive understanding of sintering and grain growth in barium titanate is yet to be achieved. Complexities arising from stoichiometry variations, defects, impurities, grain boundary and surface states, combined with material sensitivity to processing have resulted in an inability to accurately predict and control their microstructure and properties. Significant benefit can be derived from being able to understand and control the processing of these materials.

Sintering behavior is one of the most important parameters in processing barium titanate materials. Liquid phase sintering and control of microstructure is another area requiring immediate attention due to its commercial significance. While a general consensus on the sintering behavior and properties can be found in literature, there continues to exist considerable debate on trigger mechanisms responsible for abnormal grain growth in these materials. Conventional sintering studies rely on theoretical modeling and ex-situ experimental and analytical studies in predicting the behavior of ceramic compacts under high temperatures. These methods have had limited success in developing a comprehensive model of sintering with which precise control of microstructure and properties in advanced ceramics is possible. Theoretical models of sintering provide valuable insight in understanding sintering behavior but tend to be of a simplistic nature and are rarely able to account for complex behavior in multi particle systems with wide variations in particle size and distribution [1]. Analytical techniques such as dilatometry, gas adsorption, porosimetry and quantative microscopy relate sintering behavior to shrinkage, pore size and geometry. While highly instructive, the individual limitations of each technique requires careful selection of the method employed and interpretation of the subsequent results [1].

Dynamic observations of sintering can provide means of developing a deeper understanding of the high temperature processes occurring in materials. Often limitations in these studies result from the high temperatures encountered in sintering which do not permit for an intimate observation of these processes. Optical methods have been successfully used in studies on initial stages in sintering and demonstrate the advantages of such an approach [2]. Optical techniques, however lack the resolution and depth of field requirements necessary to perform in-depth studies on advanced fine ceramics. Electron microscopes, while possessing sufficient resolution capabilities, have disadvantages of high vacuum requirements. Further difficulties are encountered in imaging insulating materials. Heating materials to high temperatures result in the emission of infrared and visible light along with increased thermionic emission of electrons. Signal to noise ratios can be dramatically decreased resulting in severe image degradation [3, 4]. Above critical temperatures, imaging of the sample will no longer be possible [3]. Conventional microscopes are unable to account for interactions between the ceramic system and its surrounding atmosphere, necessitated by the high vacuum requirements of the microscope. Sintering behavior of ceramics is a function of the partial pressure of the components in the system. Mass transport and diffusion processes occurring in the system can determine the sintering behavior and hence the final sintered microstructures. Ceramic sintering behavior under conventional scanning electron microscope vacuum environments can not be expected to replicate typical furnace environments. Conventional electron microscopes have been successfully adapted to image high temperature behavior in materials [4]. However, a true understanding of ceramic sintering behavior would require experimental observation at different pressures, gases and concentrations for thorough investigations into their behavior. The dynamic capabilities of the environmental scanning electron microscope (ESEM) have been used in dynamic observations at temperatures ~1100°C and could be instrumental in studying and controlling high temperature behavior in ceramics [5-7].

Environmental scanning electron microscopy (ESEM) combines the advantages of conventional high-resolution electron microscopy with novel gas based detection, permitting observation of materials in gaseous atmospheres [8]. Gaseous secondary detectors work on the principles of operation similar to a Townsend gas capacitor [9] with gaseous ionization through collision cascade mechanisms forming the basis for signal generation and detection. In contrast with conventional detection, gaseous presence can be used to amplify signal intensity by over 1000 times [9]. Added advantages of gaseous detection stem from the ability to neutralize charge build-up on insulating samples along with the ability to vary gaseous species and chamber pressure condition [9]. This technique may be able to provide the missing link in developing an improved understanding of high temperature sintering behavior. This paper looks at the feasibility of using ESEM techniques to study the high temperature sintering behavior of barium titanate compacts which have been processed using conventional ceramic powder processing techniques.

EXPERIMENTAL PROCEDURE

Ceramic method

Undoped $BaTiO_3$ samples were prepared by ball-milling high purity $BaTiO_3$ (Ba/Ti = 1.004) powders in a 60% isopropyl alcohol / 40% deionized water solution with Darvan 821A dispersant and polyethylene glycol binder. Detailed descriptions of these procedures have been described in the literature [10]. Dried powders obtained from the slurry were pressed at 30 Ksi into disks which were then fractured prior to being used in the experiments. Binder burnout at 600°C was performed on suitable fragments which were then used in the sintering experiments.

Hot-stage detail

Sintering studies were performed using a hot stage accessory with a maximum temperature capability of 1500°C designed for use in the ESEM. The stage consists of a water-cooled aluminum base containing a heating wire used to heat a magnesium oxide crucible. The crucible is coated with platinum to ensure electrical and thermal conductivity of the setup for purposes of image optimization through voltage biasing while also maintaining uniform temperature characteristics within the crucible. Furnace power and programming were controlled externally through the microscope controller.

Microscope detail

The microscope used in these experiments was a FEI XL-30 environmental scanning electron microscope (ESEM) with a field emission source and a theoretical resolution of 1.5nm. Gas-based detection employed in the ESEM permits operation of the instrument in varying gaseous environments in pressures ranging between 1-20 torr depending on the choice of detector. A ceramic gaseous secondary detector (GSED) with a wire hook adaptor was used in these experiments. The chamber was protected from the high temperatures by means of a heat-shield attachment which also acted as a voltage bias in conjunction with the Pt-coated crucible.

Procedure

The furnace setup was assembled and tested in the ESEM. Samples were loaded in the furnace and the chamber was pumped down in the gaseous imaging mode. A five step pump-down sequence was employed with the chamber being cycled between 1 and 10 torr. Water vapor was used as the imaging gas with nitrogen being used as the purge gas. The furnace controller was programmed to study a conventional sintering cycle. All experiments were commenced at 500°C as the changes in microstructure were not observed below this temperature. A heating rate of 10°C/min was employed to attain temperature values between 1320-1375°C. This was followed by a two hour dwell at the set sintering temperature. Cooling rates of 15 °C/min down to 800°C and 20°C/min to room temperature were employed to mimic furnace cooling conditions. A chamber pressure ~ 2 torr was used in all the experiments. A 30KV beam voltage was used to obtain images of the changes occurring in the sample. A combination of sample bias voltage, shield voltage and beam spot-size and chamber gas pressure was used in improving the image quality through the sintering cycle. Digital images were obtained as the cycle was executed to completion.

RESULTS AND DISCUSSIONS

The following observations can be made based on comparisons of the sintering studies which have been carried out in the ESEM under the above mentioned conditions. Figure 1(a) shows the morphology of the milled and dried powders. A fairly homogenous distribution of particles (~ 0.5 - 1μ) can be seen. Figure 1(b) shows the morphology of the green compact after binder burnout at 600°C, prior to sintering in the ESEM. Clear differences can be seen with Figure 1(b) showing the closer packing of particles. Further initial stage inter-particle contact and necking has also occurred with a slight increase in particle size.

As the temperature is raised from 500°C to 1100°C, no noticeable changes are observed in the sample. Continuous adjustments of the imaging parameters are necessary to maintain optimum imaging conditions as thermionic emission from heating the sample results in increasing signal noise. Adjustment of contrast, brightness, sample voltage, shield voltage and spot size parameter need to be made to maintain adequate imaging contrast. Between 1100-1200°C, a distinct and irreversible transformation is observed in the powder compacts with the formation of two dimensional needle shaped structures. This transformation is observed in all the in-situ experiments carried out in this study. Figure 2 shows the onset of this transformation. It was found that this transformation results in a general degradation of imaging conditions. Signal collection and amplification in gaseous detectors are sensitive to fluctuations in chamber pressure, and it is likely that the evolution of gaseous species during the transformation may have resulted in the pressure fluctuations causing the image disturbances before returning to equilibrium. Thermogravimetric studies in these temperature ranges have shown the evolution of CO_2 in barium titanate as a result of decomposition of carbonate impurities [11]. Continual heating to sintering temperatures between 1320°C and 1375°C results in growth of these structures into hexagonal plates with their final sizes directly dependent on the sintering temperature. Figures 3 (a) to (d) have been taken at different stages in the microstructural evolution of a sample sintered at 1350°C. Beginning with primary particles ranging in size from 0.5-1μ (Figure 3(a)) solid state processes occurring in the sample result in sintering and grain growth of the particles with the final grain-sizes in the 3-6μ range. It must be stated that solid state controlled mass transport and diffusion mechanism appeared to be predominantly responsible for sintering based on the relatively slow growth rates and refined final grain-size of the final microstructures. It is important to observe that the sintering behavior in these studies favor anisotropic growth of hexagonal plate-like grains. Sintering barium titanate under reducing oxygen deficient conditions has been shown to cause formation of hexagonal barium titanate [12-14]. While it was not possible to measure oxygen partial pressures in this study, it is expected that sintering conditions resulting from the use of water vapor as the imaging gas along with an N_2 purge gas and pressures in the ~2 torr range resulted in a highly reducing sintering atmosphere. These conditions are also expected to be the cause of the small grained microstructures in this study, as identical materials sintered in atmospheric conditions were shown to have dense large grained microstructures (~60μ) [10,15,16]. Final in-situ sintered microstructures developed in samples sintered at 1320°C and 1375°C are shown in figure 4 (a) and (b). Distinct differences in final grain size are observable with the 1375°C sample having a larger final microstructure.

Typical Ba-excess materials exhibit solid state sintering mechanisms characterized by high densities and small grained microstructures with linear dependence of grain size on the sintering temperature. Studies in our group show deviations from this behavior with Ba-excess powders (Ba/Ti$\leq$ 1.006) exhibiting atypical sintering behavior and abnormal grain growth characteristics. It has been proposed that variations in local stoichiometry can result in the formation of discrete Ti-excess liquid phase formation which is responsible for abnormal growth mechanisms [15,16]. Formation of liquid phase regions can be seen in Figures 5 (a) and (b) of in-situ sintered samples at 1350°C and 1375°C. Inhomogeneous distribution of the liquid is clearly visible. These microstructures appear in agreement with the above mentioned theories. It must be noted that the lack of abnormal growth observed in these microstructures is attributed to the reducing conditions present during sintering in these samples.

CONCLUSIONS

Sintering behavior of undoped barium titanate powders (Ba/Ti ratio 1.004) has been studied in the ESEM at temperatures between 1320°C and 1375°C. Solid state sintering mechanisms can be seen to dominate in the current experiments. Anisotropic grain growth resulting in hexagonal plate-like grains is observed. Reducing conditions resulting from low pressures and oxygen deficient sintering environment appear to be responsible for this behavior. Discontinuous liquid phase region are found in samples sintered at 1350°C and 1375°C, substantiating previous work in understanding abnormal growth mechanisms in these materials. These studies demonstrate the capabilities of the ESEM technique in studying high temperature properties of ceramic materials. Conventional sintering studies in controlled low pressure environments are currently in progress and will be helpful in understanding the high temperature ESEM investigations carried out in this work. Additional work in expanding this research to observe the behavior of barium titanate under conditions of different pressures, and gaseous environments in the ESEM needs to be performed.

REFERENCES

[1]H.E. Exner, "Solid-State Sintering: Critical Assessment of Theoretical Concepts and Experimental Methods," Powder Metallurgy, **4** 203-209 (1980).

[2]W.D. Kingery and M. Berg, "Study of the Initial Stages of Sintering by Viscous Flow, Evaporation-Condensation, and Self-Diffusion," Journal of Applied Physics, **26**(10) 1205-1212 (1955).

[3]J. Edelmann, "Thermal Experiments"; pp. 96-109 in *In situ Scanning Electron Microscopy in Materials Research* 1st Edited by K. Wetzig and D. Schulze. Akad. Verl. GmbH , Berlin, 1995.

[4]G. Gregori, H.-J. Kleebe, F. Sieglin and G. Ziegler, "*In situ* SEM Imaging at Temperatures as High as 1450°C," Journal of Electron Microscopy, **51**(6) 347-352 (2002).

[5]N.S. Srinivasan, "Dynamic Study of Changes in Structure and Morphology during the Heating and Sintering of Iron Powder," Powder Technology, **124** 40-44 (2002).

[6]E.R. Prack, "An Introduction to Process Visualization Capabilities and Considerations in the Environmental Scanning Electron Microscope (ESEM)," Microscopy Research and Technique, **25** 487-492 (1993).

[7]P.W. Brown, J.R. Hellmann, and M. Klimliewicz, "Examples of Evolution in Ceramics and Composites," Microscopy Research and Technique, **25** 474-486 (1993).

[8]G.D. Danilatos, "Introduction to the ESEM instrument," Microscopy Research and Technique, **25** 354-361, (1993).

[9]A.L. Fletcher, B.L. Thiel and A.M. Donald, "Amplification Measurements of Alternating Imaging Gases in the Environmental SEM," J. Phys. D: Appl. Phys., **30** 2249-2257 (1997).

[10]S. Subramaniam, "Effects of Processing on PTCR Barium Titanate Systems with Barium Oxide and Titanium Oxide Additions in the Near-stoichiometric region", M.S. Thesis. University of Cincinnati, Cincinnati, Ohio, 2002.

[11]M. Demartin, C. Herard, C. Carry and J Lemaitre, "Dedensification and Anomalous Grain Growth during Sintering of Undoped Barium Titanate," J. Am. Ceram. Soc., **80**(5) 1079-1084 (1997).

[12]D. Kolar, U. Kunaver, A. Recnik, "Exaggerated Anistropic Grain Growth in Hexagonal Barium Titanate Ceramics," Phys. Stat. Sol. (a), **166** 219-230 (1998)

[13]A. Recnik and D. Kolar, "Exaggerated Growth of Hexagonal Barium Titanate under Reducing Sintering Conditions," J. Am. Ceram. Soc., **79**(4) 1015-1018 (1996).

[14]D.C. Sinclair, J.M.S. Skakle, F.D. Morrison, R.I. Smith and T.P. Beales, " Structure and Electrical Properties of Oxygen-deficient Hexagonal $BaTiO_3$," J. Mater. Chem, **9** 1327-1331 (1999).

[15]G. Liu, R.R. Roseman, "Effect of BaO and SiO_2 additions on PTCR $BaTiO_3$ Ceramics," J. Matl. Sci. 341-347 (1999).

[16]N. Mukherjee, R.D. Roseman, Q. Zhang, "Sintering Behavior and PTCR Properties of Stoichiometric Blend $BaTiO_3$," J. Phys. Chem. Sol., **23** 631-638 (2001).

Fig. 1: Powder morphologies (a) milled powders (b) green compact after 600°C binder burnout

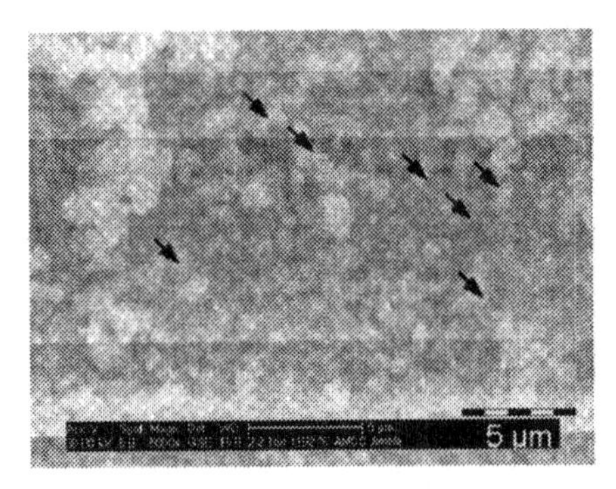

Fig. 2: Nucleation of needle-like structures at 1192°C in sample sintered at 1375°C

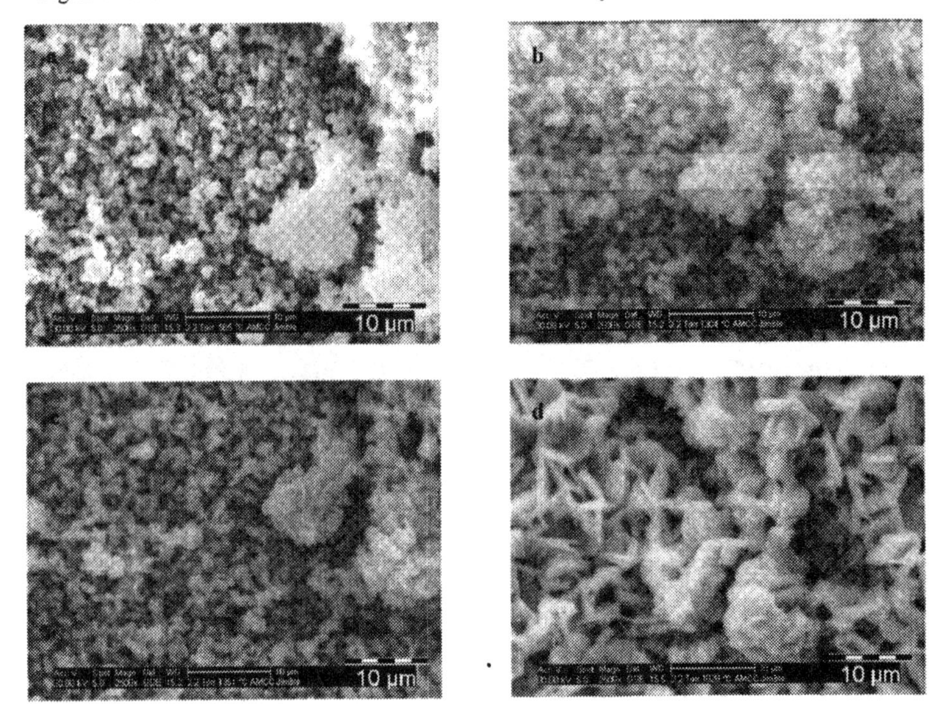

Fig. 3: In situ observation of microstructure evolution in specimen sintered at 1350°C
(a) 565°C heating
(b) 1304°C heating
(c) 1350°C dwell after 1 minute
(d) 1028°C cooling

Fig. 4: Comparison of final microstructures of sintered samples (a) 1320°C and (b)1375°C

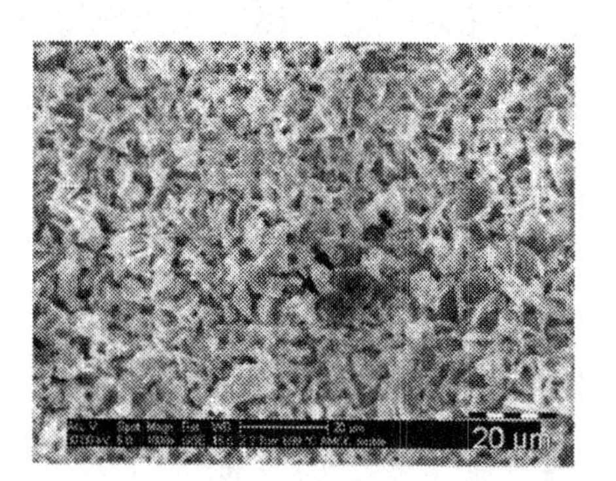

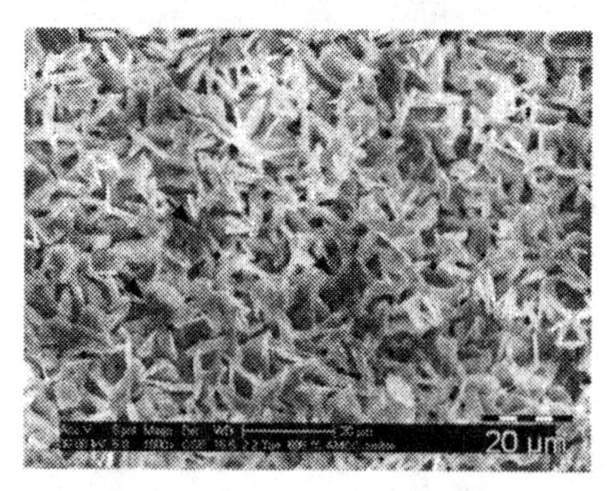

Fig. 5: Inhomogenous liquid phase presence in samples sintered at 1350°C (a) and 1375°C (b)

A MESOSCALE DESCRIPTION OF MICROSTRUCTURAL EVOLUTION FOR SLIP CAST ALUMINA SINTERED AT 1350°C

R.J. McAfee Jr. and I. Nettleship,
Department of Materials Science and Engineering,
University of Pittsburgh,
Pittsburgh, PA 15261.

ABSTRACT

This study used a simple pore boundary tessellation to achieve a mesoscale description of microstructure evolution for intermediate and final stage sintering of slip cast alumina. Cell maps can be constructed by plotting the distribution of one tessellation cell property against another. Densification below a solid volume fraction of 0.9 was dominated by a reduction in the frequency of low solid area fraction tessellation cells containing large pore sections and an order of magnitude decrease in pore section area associated with the high frequencies in the cell maps. Densification above a solid area fraction of 0.9 was associated with an increase in the size of the high solid area fraction cells that can be interpreted in terms of pore shrinkage and elimination.

INTRODUCTION

The description of microstructural evolution for the sintering of ceramics has been developed almost exclusively at the grain-scale with parameters such as grain size and pore size. This trend is reinforced by the traditional phenomenological models that are based on representative volumes that contain only one grain and its attendant porosity. In the models, the microstructure evolves by contact flattening and uniform reductions in the pore radius [1][2]. Subsequently these models have been combined with self-similar coarsening laws to include the effect of grain growth [3][4]. In general, the experimental studies have shown that grain growth occurs mainly at higher solid volume fractions [5] and the pore intercept remains constant through intermediate stage sintering [6][7]. There are also some rarer experimental studies that have reported how the grain size distribution evolves [8][9]. Mercury intrusion porosimetry remains a commonly used method for determining the evolution of the pore size distribution during sintering [10][11]. Unfortunately this method depends on the cylindrical pore shape assumption and is known to underestimate the pore size, especially for larger interior pores [12]. It is also unable to generate information about the solid phase or the spatial distribution of the phases. It can therefore be concluded that the quantitative microstructural measurement methods used in the past have been limited to the grain-scale and therefore do not directly address higher microstructural scales associated with particle packing behavior.

Many studies have used qualitative microstructural observations to study the effect of spatial distribution on sintering, particularly with regard to the effects of agglomeration and mixing powders of different size [13][14][15]. It was clear from these observations that the microstructures were heterogeneous at scales higher than the grain-scale and grain rearrangement was required to reach full density [15]. The implication is that quantitative microstructural studies must be done at scales above the grain-scale to directly observe the manifestations of microstructural heterogeneity. A limited number of simulations [16][17] have gone beyond elements that contain only one grain to accommodate the qualitative observations of microstructural heterogeneity in sintering mentioned above. Although these simulations were two dimensional and quite dependent on the initial assumed arrangement of particles they did reproduce the granular mechanisms of rearrangement, neck rupture, grain rotation and differential sintering that have been observed in in-situ experiments [18][19][20] or inferred from interrupted experiments [15]. However, there are no established experimental techniques for quantifying spatial changes in the arrangement of grains or pores during the microstructural evolution in sintering.

In order to examine sintering microstructures at a higher scale it is necessary to partition the microstructure with regard to features other than single grains. One might consider looking at the local spatial arrangements by partitioning sintering microstructures based on pores or clusters of grains. Either would be difficult in circumstances where the chosen microstructural feature is present in high solid volume fractions [21]. If this is the case, tessellation is perhaps the simplest way of partitioning space in a unique fashion. Tessellation has been used to model the mechanical behavior of heterogeneous composite materials with a low volume fraction second phase [22][23], permeability of heterogeneous porous materials [24] and the sintering of hard metals [25]. Tessellation has also been used to characterize reinforcement distributions in composites [26]. The present experimental study used a simple pore boundary tessellation to examine the spatial arrangement of pores. It is thought that this will allow new descriptions of microstructural evolution in sintering that reflect the manifestations of spatial heterogeneity in particle packing.

EXPERIMENTAL PROCEDURES

A 50 g specimen was prepared from alumina powder with an average particle size of 0.3μm (Premalox, Alcoa, Pittsburgh). First the powder was dispersed in deionized water with an ammonium polyacrylate dispersant (Darvan C, Vanderbilt) and then cast in a plaster mold to give a green density of 59%. The sample was then sintered for 0.1 hour at 1350°C and cut into smaller samples. The densities of these pieces were tested using the Archimedes method in which the samples were evacuated before immersing them in water. This showed that the solid volume fraction of the samples did not vary by more than 2% around an average of 0.82. Then the samples were heat treated at the same temperature for times up to 60 hours. Sections through the samples were ground and polished before thermal etching at 1325°C for 18 minutes. The densities of the samples were measured before and after the etching treatment to make sure that the thermal etching did not change the solid volume fraction of the samples.

The surfaces of the samples were observed by SEM (Philips XL30 FEG) and images were taken at an appropriate magnification. The images were processed to give pore section images and grain boundary reconstructed images. The volume fraction of pores was then determined from the pore images and the 95% confidence interval calculated from the variance between the images for each sample. The value of the volume fraction of pores from the Archimedes method was within the confidence intervals for the results of the image analysis. The images were therefore taken to be collectively representative of the material. The pore images were also used to determine the distribution in pore areas and the distribution in the pore shape parameter. Finally the pore boundaries of the pore images were tessellated to partition the microstructure into an array of space filling cells. Figure 1 shows an illustration of a pore boundary tessellated structure superimposed on a micrograph. The tessellation cell includes the pore and portions of the surrounding grains and is therefore sensitive to the local packing arrangement. The properties of the cells that did not intersect the edge of the images were calculated. The populations of the cell properties such as solid area fraction of the cell (SAF), cell area (CA), pore section area (PA) and cell coordination number were then calculated.

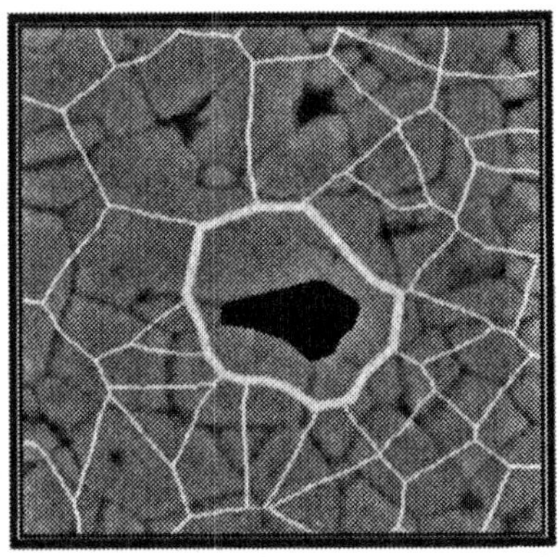

Figure 1: A SEM micrograph of a microstructure with superimposed tessellations cells.

RESULTS

The solid volume fraction increased with time at 1350°C from 0.82 at 0.1 hours to 0.96 after 15 hours in accordance with the typical semi-log relationship [27]. Beyond this time a limiting density of 0.97 was reached. Figure 2 shows the distributions in pore areas measured from the pore images. This is expressed in terms of the fractional area of the entire image covered with pores in that size interval. At the shorter sintering times a smooth unimodal distribution existed between 0.01μm^2 and 5μm^2 which is somewhat skewed towards the larger sizes with the peak at approximately 0.3μm^2. At this point in the densification the average grain size is approximately 0.4μm [30]. As the sintering continues the pore area distributions remain unimodal and, the peak in the distribution remained in the range 0.01 to 0.03μm^2, although it occurs at much lower fractional areas. The peripheries of the distributions are relatively unchanged so that the distribution widens as the sintering time increased. The solid area fraction calculated from micrographs was in good agreement with the Archimedes measurements of solid volume fraction. It is therefore concluded that there was no significant pore area at sizes below the resolution of the measurements.

An example of a cell map distribution derived from the pore boundary tessellations of the samples sintered for 0.1hours is shown in figure. 3. In this case the axes in the base plane are the solid area fraction of the cell (SAF) and the area of the cell (CA). This sample had a solid volume fraction of 0.87 and it is clear that at the scale of the tessellation, defined by the area of the cells, the sample is quite heterogeneous. The cells have a wide range in solid area fraction extending from 0.6 to 0.99. The evolution in the solid area fraction (SAF) and cell area (CA) maps in the first 15 hours can be seen in figure 4. Careful examination of the contour maps shows that densification within the first hour caused a decrease in the frequency (grayscale gets lighter) of low solid area fraction cells and an increase in the frequency (grayscale gets darker) of cells with solid area fractions above 0.9. This is more obvious for the intervals with higher frequencies (peak in the distribution) which indicate a marked increase in SAF and a slight decrease in CA. For 10 hours and beyond the cell areas of the high SAF cells begin to increase. Further information can also be obtained by plotting cell maps of cell area (CA) against the area of the pore contained in that cell (PA). Contour maps for the evolution of these distributions are shown in figure 5. Again the distributions appear wide. It is particularly noteworthy that as the sintering time increases the higher frequencies (dark intervals) shift to smaller pore sizes.

Summarizing the changes in the high frequency intervals in the cell maps, the sintering of the alumina resulted in a decrease in the frequency of low solid area fraction cells containing large pores up to a solid volume fractions of approximately 0.9 and then a marked increase in cell size above solid volume fractions of 0.9.

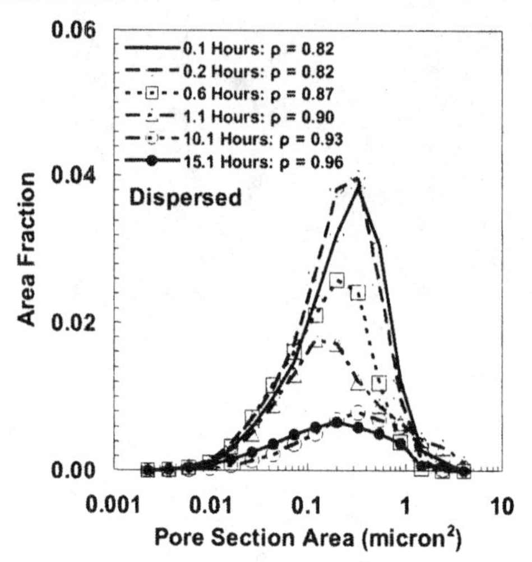

Figure 2: The evolution of the pores section area distribution during sintering.

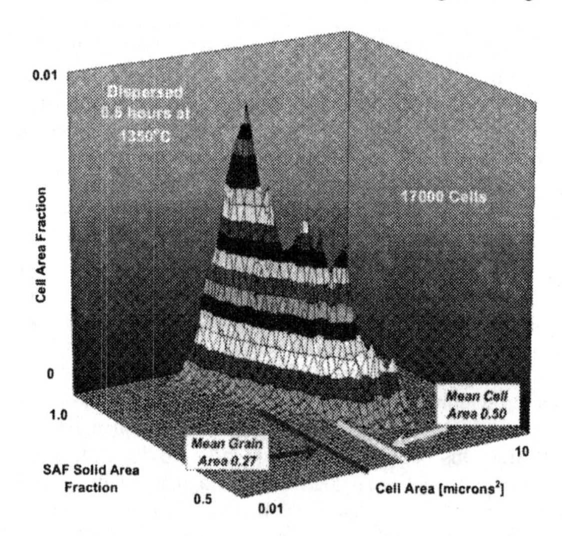

Figure 3: Cell map showing the correlation between the solid area fraction (SAF) and the cell area (CA) for the tessellation of the sample sintered for 0.5 hours.

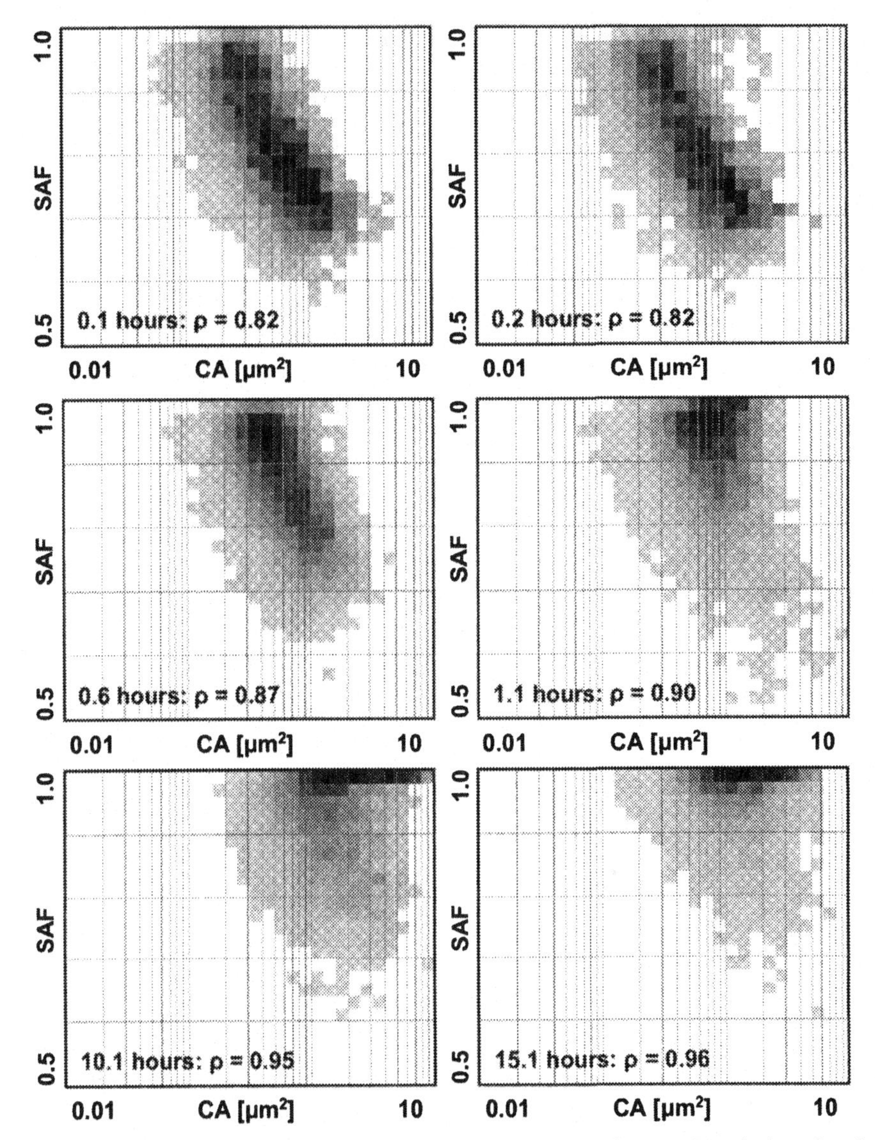

Figure 4: Contour cell maps of solid area fraction (SAF) vs. cell area (CA) during sintering. Each grayscaled square represents an interval. The grey scale represents the relative frequency with white representing low frequency and black representing high frequency.

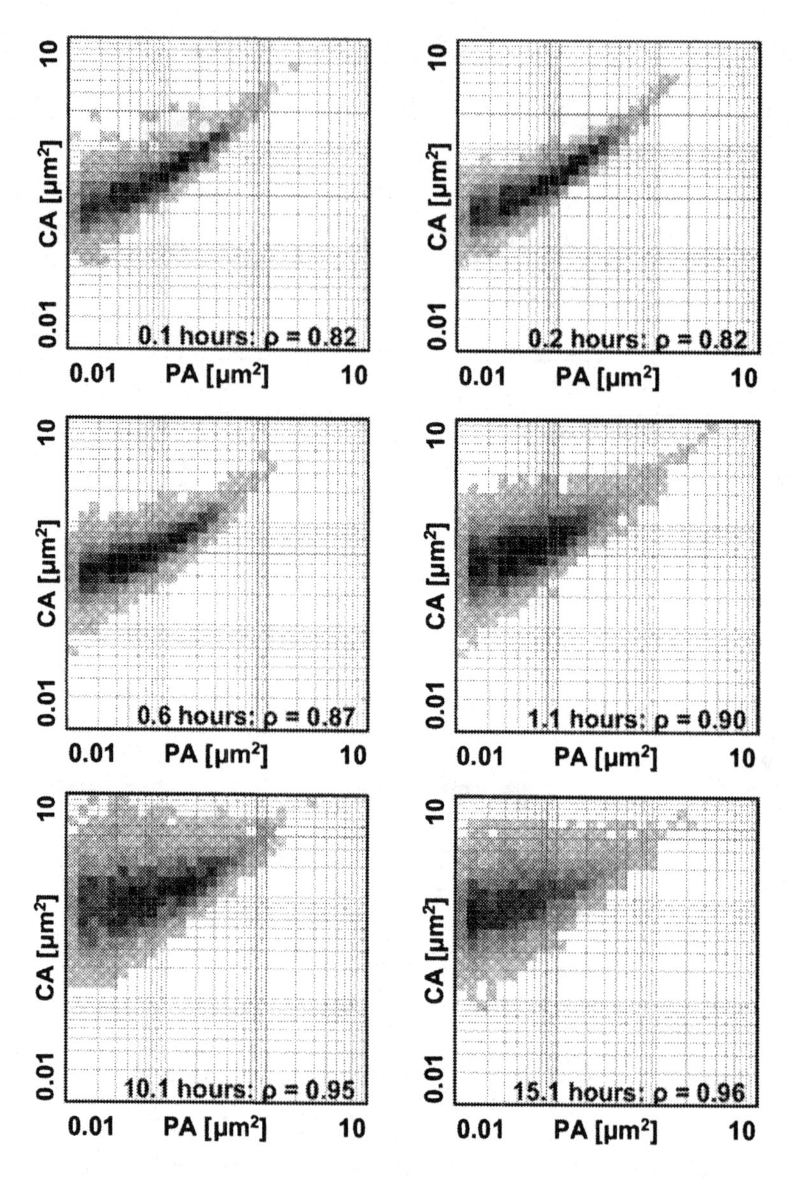

Figure 5: The evolution in the contour cell maps of cell area (CA) vs. pore area (PA) during sintering.

DISCUSSION

The descriptions of microstructural evolution in figures 2, 4, and 5 are based on the pores rather than the grains. The results of the pore area distributions presented in figure 2 suggest that for the first hour, the average pore area does not change much as sintering continues. In contrast, the overall area of the surface occupied by pores in each area size class decreases quite markedly in the middle of the distribution. This is consistent with a number of studies on ceramics and powder metals [6][7][28][29] that showed that the average pore intercept length does not change through intermediate stage sintering. This point has been appreciated for some time although much of the phenomenological modeling of intermediate stage sintering persists with the assumption that shrinkage occurs solely by contact flattening and uniform shrinkage of pores along with the radius of their sections. If this reflected the major changes in the pore structure over intermediate stage densification then one would expect that the average pore intercept would decrease markedly and the distribution in pore areas would shift to smaller sizes. This is clearly not the case, in fact the results for the material in this study suggest that the very largest pore sections are not shrinking or being eliminated and the frequencies of the smallest pores are also unchanged. In contrast, the pore areas in the intermediate size intervals decrease with the fastest decrease taking place about the average, causing a broadening of the distribution. One potential explanation is that these pores are collapsing or breaking up by granular mechanisms involving grain rearrangement. The rearrangement processes may require that grain contacts be ruptured followed by grain boundary sliding and new contact area formed throughout intermediate stage sintering. While there is evidence of neck rupture [15] and grain rearrangement during in-situ experiments [19] they are not addressed in the phenomenological models of sintering beyond densities corresponding to dense random packing [11][30]. One key consequence of such a pore break up mechanism would be an increase in the number of pore sections per unit area which is not observed in this study. In fact the number of pore sections per unit area decreases as the microstructure evolves. The cell maps derived from the pore boundary tessellations in figures 4 and 5 give some extra information pertaining to the microstructure evolution on a scale that is sensitive to the local arrangement of pores. This can be used to further speculate on the nature of the observed microstructural evolution. As pointed out earlier, the microstructure is quite heterogeneous on this scale and the evolution at solid volume fractions below 0.9 can be interpreted as the reduction in the frequency of low SAF cells that contain relatively large pores. In fact the pore section areas associated with the high frequencies in the cell map decreases by an order of magnitude during intermediate stage sintering. There is also a reduction in the cell size suggesting that the large pores could be breaking up. However pore shrinkage and elimination of the smaller and much more numerous pore sections may lead to the observed overall reduction in the number of pores per unit area. Pore shrinkage and elimination would tend to increase the average cell size therefore reducing the effect of large pore break up on the cell area distribution. Above a solid volume fraction of 0.9 the peak in the SAF vs. CA moves to larger CA and higher SAF while maintaining constant pore section area. This is broadly consistent with pore shrinkage and elimination as would be expected of microstructure evolution during final stage sintering.

It is clear that the cell map description gives a basis for a new quantitative understanding of mesoscale microstructural evolution of densification in which granular mechanisms of grain rearrangement may play a key role. This description is also consistent with the findings of in-situ experiments [19] in that grain rearrangement processes occur at high solid volume fraction and must therefore be cooperative in nature. Obviously, time resolved in-situ experiments would be necessary to probe the nature of these mechanisms in fine-grained ceramics. However, the tessellation based method used in this study is a relatively simple way of inferring aspects of this behavior based on its manifestations in the tessellation cell population properties during microstructural evolution.

The question may arise as to the utility of this inherently two dimensional tessellation procedure. However, it must be remembered that this is not a two dimensional model simulation of microstructural evolution. On the contrary, it is a method of partitioning a plane with regard to the spatial arrangement of features that intersect the plane in a real material. The microstructural processes that resulted in this arrangement are occurring in three dimensions. While it is necessary to keep this in mind for interpretation of microstructural evolution, the conclusions that are inferred are valid if the plane is random and the images representative. In this sense the basic assumption is no different than that used to measure the grain size distribution. The tessellation itself is a unique way of partitioning the plane in terms of the spatial distribution of pores reflecting the way that they intersect the plane. Therefore the cell map distributions are properties of sintering microstructures in the same sense as grain section populations. Indeed, if the pores are fully connected, as in the commonly accepted interpretation of intermediate stage sintering, the three dimensional tessellations of the pore boundaries could not be used to partition the microstructure in a meaningful way. Like all other measurements taken from a section the cell maps may be easily interpreted but they are not capable of providing topological information such as connectivity in three dimensions. It is therefore not possible from this measurement alone to determine for example, if individual pores are isolated or connected.

CONCLUSIONS

The tessellation of pore boundaries on polished sections has been used to generate a mesoscale description of the sintering of alumina. The properties of the cells reflected the characteristics of the pore sections and their immediate surroundings. At solid volume fractions below 0.9 the major microstructural change at this scale involved a decrease in the frequency of low SAF cells containing large pore sections. The pore section area associated with the high frequencies in the SAF vs. CA maps decreases by an order of magnitude. This is thought to be consistent with the effects of cooperative grain rearrangement. Above a solid volume fraction of 0.9 the major change in the cell population involves an increase in the frequency of large, high solid area fraction cells possibly due to the pore shrinkage and elimination that would be expected in final stage sintering.

ACKNOWLEDGEMENTS

The authors would like to thank the National Science Foundation for financial support under the award DMI-980430.

REFERENCES

1. W.D. Kingery and M. Berg, "Study of the Initial Stages of Sintering Solids by Viscous Flow, Evaporation -Condensation and Self Diffusion," *J. Appl. Phys.* **26** 1205-1212 (1955).

2. R.L. Coble, "Sintering Crystalline Solids. I Intermediate and Final State Diffusion Models," *J. Appl. Phys.*, **32** 787-792 (1961).

3. C.A. Handwerker, R.M. Cannon and R.L. Coble, "Final Stage Sintering of MgO," pp. 619-643, Advances in Ceramics 10, *Structure and Properties of MgO and Al_2O_3*, edited by W.D. Kingery, The American Ceramics Society, Columbus, OH (1984).
4. M.P. Harmer, "Use of Solid-Solution Additives in Ceramics Processing," pp. 679-696, Advances in Ceramics 10, *Structure and Properties of MgO and Al_2O_3*, edited by W.D. Kingery, The American Ceramics Society, Columbus, OH (1984).
5. C. P. Cameron and R. Raj, "Grain-Growth Transition During Sintering of Colloidally Prepared Alumina Powder Compacts," *J. Am. Ceram. Soc.* **71** 1031-35 (1988).
6. N.J. Shaw and R.J. Brook, "Structure and Grain Coarsening During the Sintering of Alumina," *J. Am. Ceram. Soc.*, **69**107-110 (1986).
7. M.D. Lehigh and I. Nettleship, "Microstructural Evolution of Porous Alumina," *Materials. Research. Society. Symposium. Proceedings*, **371** 315-320 (1995).
8. J.M. Ting and R.Y. Lin, "Effect of Particle Size Distribution on Sintering, Part II Sintering of Alumina," *J. Mat. Sci.*, **30** 1882-1889 (1995).,
9. I. Nettleship, R. J. McAfee and W. S. Slaughter, "The Evolution of the Grain Size Distribution During the Sintering of Alumina at 1350°C," *J. Am. Ceram. Soc.*, **85** 1954-1960 (2002).
10. J. Zheng and J.S. Reed, "Effect of Particle Packing Characteristics on Solid-State Sintering," *J. Am. Ceram. Soc.*, **72** 810-17 (1989)
11. P.L. Chen and I.W. Chen, "Sintering of Fine Oxide Powders: I, Microstructural Evolution," *J. Am. Ceram. Soc.*, **79** 3129-41 (1996).
12. H.D. Lee, "Validity of Using Mercury Porosimetry to Characterize the Pore structures of Ceramic Green Compacts," *J. Am. Ceram. Soc.*, **73** 2309-15 (1990).
13. J.P. Smith and G.L. Messing, "Sintering of Bimodally Distributed Alumina Powders," *J. Am. Ceram. Soc.*, **67** 238-242 (1984).
14. F.F. Lange and M. Metcalf, "Processing Related Fracture Origins: II Agglomerate Motion and Crack-Like Internal Surfaces Caused by Differential Sintering, *J. Am. Ceram. Soc.*, **66**, 398-406 (1983).
15. O. Sudre and F.F. Lange, "Effect of Inclusions on Densification: I, Microstructural Development in an Al_2O_3 Matrix Containing a High Volume Fraction of ZrO_2 Inclusions, *J. Am. Ceram. Soc.*, **75**, 519-24 (1992).
16. A Jagota and P.R. Dawson, "Micromechanical Modeling of Powder Compacts-II. Truss Formulation of Discrete Packings," *Acta Metall.*, **36** 2563–2573 (1988).
17. L.R. Madhavrao and R. Rajagopalan, "Monte Carlo Simulations for Sintering of Particle Aggregates," *J. Mater. Res.*, **4** 1251–1256 (1989).
18. H. E. Exner, "Principles of Single Phase Sintering," *Rev. Powder Metall Phys Ceram.*, **1** 1–251 (1979).
19. M.W. Weiser and L.C. De Jonghe, "Rearrangement During Sintering in Two-Dimensional Arrays," *J. Am. Ceram. Soc.*, **69** 822–826 (1986).
20. M. Kusunoki, K. Yonemitsu, Y. Sasaki and Y. Kubo, "In-Situ Observation of Zirconia Particles at 1200°C by High Resolution Microscopy," *J. Am. Ceram. Soc.* **76** 763–765 (1993).
21. J.E. Spowart, B. Maruyama amd D.B. Miracle, "Multi-scale Characterization of Spatially Heterogeneous Systems: Implications for Discontinously Reinforced Metal-Matrix Composite Microstructures," *Mat. Sci. & Eng.*, **A307** 51-66 (2001).
22.K. Lee and S. Gosh, "Microstructure Based Numerical Method for Constitutive Modeling of Composite and Porous Materials," *Mat. Sci. & Eng.*, **A272** 120-133 (1999).

23. S. Moorthy and S. Ghosh, "Adaptivity and Convergence in the Voronoi Cell Finite Element Model for Analyzing Heterogeneous Materials," *Computer Methods in Applied Mechanics and Engineering*, **185** 37-74 (2000).
24. N.A. Vrettos, H. Imakoma and M. Okazaki, "Effective Medium Approximation of 3-D Voronoi Networks," *J. Appl. Phys.*, **67** 3249-53 (1990).
25. L. Mahler and K. Runesson, "Modelling of Solid-Phase Sintering of Hard Metal Using Mesomechanics Approach," Mechanics of Cohesive Frictional Materials," **5** 653-671 (2000).
26. W.A. Spitzig, J.F. Kelly and O. Richmond, "Quantitative Characterization of Second-Phase Populations," *Metallography*, **18** 235-261 (1985).
27. I. Nettleship and R.J. McAfee, "Microstructural Pathways for the Densification of Slip Cast Alumina," *Mat. Sci. & Eng.*, **A352** 287-93 (2003).
28. E.H. Aigeltinger and H.E. Exner, "Stereological Characterization of the Interaction Between Interfaces and its Application to the Sintering Process," *Met. Trans.*, **8A** 421-424 (1977).
29. E.H. Aigeltinger and R.T. DeHoff, "Quantitative Determination of Topological and Metric Properties During Sintering of Copper," *Met. Trans.*, **6A** 1853-1862 (1975).
30. H. Reidel and B. Blug, "A Comprehensive Model for the Solid State Sintering and its Application to Silicon Carbide," pp. 49-70, in *Multiscale Deformation and Fracture in Materials and Structures*, the James R. Rice 60th Anniversary Volume. Edited by T.J. Chuang and J.W. Rudnicki. Klewer Academic Publishers, Boston (2001).

Engineering Ceramic Processes and Microstructures

COUPLED DIFFUSION COEFFICIENTS FOR MASS TRANSPORT OF COMPOUNDS

Dennis W. Readey
Colorado Center for Advanced Ceramics
Dept. of Metallurgical and Materials Engineering
Colorado School of Mines
Golden, CO 80401

ABSTRACT

For mass transport of compounds, all of the constituent chemical species must be transported. This leads to coupling of their fluxes to maintain either stoichiometry or electrical neutrality. This coupling gives an effective or apparent transport coefficient that is a function of the individual transport coefficients of the species transported. The cases of coupled: solid-state diffusion driven by curvature; gas transport driven by curvature; and active gas corrosion demonstrate the important considerations and give different effective transport coefficients for each process.

INTRODUCTION

A number of years ago, an effort was made to clarify the effects of dopants and curvature on the defect equilibria and mass transport in ionic solids during sintering.[1,2] In spite of the clear demonstration that, for point defect chemical equations (Kröger-Vink[3] notation) such as

$$O_O^X = \tfrac{1}{2}O_2(g) + V_O^{\bullet\bullet} + 2e' \tag{1}$$

leading to the equilibrium constant

$$\frac{[V_O^{\bullet\bullet}][e']^2 p_{O_2}^{\frac{1}{2}}}{[O_O^X]} = K, \tag{2}$$

the square brackets, [ ], are atomic site fractions and fraction of occupied states in the conduction band, $[e']$, as required by statistical thermodynamics.[1,2,4] And the correct electrical neutrality is: $2 \times e \times n_{V_O^{\bullet\bullet}} = n \times e$ where $n_{V_O^{\bullet\bullet}}$ = the number of vacant oxygen sites (cc^{-1}), n = the number of electrons (cc^{-1}), and e is the electron charge (C). More often than not, the incorrect expression $2[V_O^{\bullet\bullet}] = n$ is taken as electrical neutrality, which it clearly is not, since $[V_O^{\bullet\bullet}]$ is a fraction and has no charge.[5] At best, this leads to inconsistency in point defect equilbria in the literature and, at worst, errors in quantitative predictions of "concentrations" that can be as large as several orders of magnitude. This will not be addressed further here.

However, what is addressed here, also part of the earlier results,[1,2] is related and is the inconsistency in expressions for ambipolar diffusion coefficients in the literature.[6,7] This confusion has arisen because flux expressions usually are not related to gradients of measurable chemical potentials, which can be the result of curvature in sintering and coarsening processes.[6,7] Therefore, it is shown how correct expressions for ambipolar solid-state diffusion coefficients under general chemical potential gradients, which may be determined by curvature, can be obtained with a slightly different approach than before.[1,2] In addition, the coupling of diffusion fluxes for compounds is extended to gas diffusion important in active gas corrosion and transport processes due to curvature including coarsening, grain boundary grooving, and neck growth during sintering where vapor pressures are high enough.

AMBIPOLAR SOLID STATE DIFFUSION AND MASS TRANSPORT

Gradient and Ambipolar Diffusion

For the sake of concreteness, assume that the chemical potential gradient is determined by the neck curvature between particles as in the simple 2-sphere model of sintering, Figure 1.[8,9] Also for the sake of concreteness and simplicity, assume a binary ionic compound such as MgO. A more general compound such as $A_\alpha B_\beta$ could be considered but this makes the algebra more complex and results less transparent.

From the neck between the particles to the center of the particles of Figure 1, there is a chemical potential gradient determined by the neck curvature, κ, (neglecting the curvature effect of the particle radius, r):

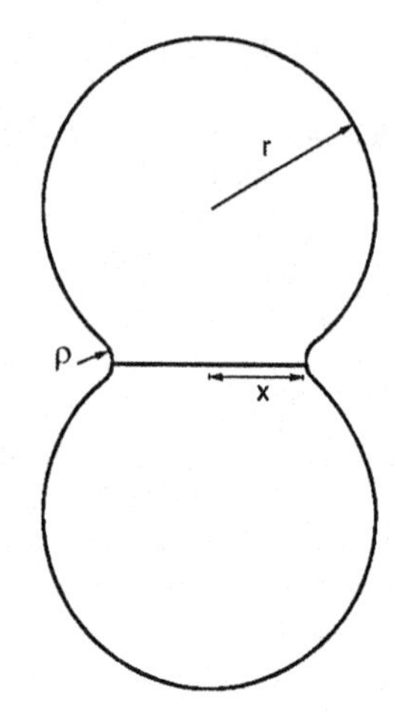

Figure 1. Two-sphere sintering model

$$\nabla \mu_{MgO} = g\kappa\gamma\Omega \tag{3}$$

where μ_{MgO} = chemical potential of MgO (J/mole), γ = surface energy (J/m^2), g = geometric factor, κ = neck curvature ≈ $-1/\rho$, ρ = neck radius of curvature (m), and Ω = molar volume (m^3/mole). Then, the flux of MgO, J_{MgO} (mole/m^2-s), is given by:

$$J_{MgO} = J_O = J_{Mg} = -C_{Mg} B_{amb} \nabla \mu_{MgO} \tag{4}$$

where $C_{Mg} = C_O = C_{MgO}$ = concentration of magnesium ions ("moles"/m^3) and B_{amb} = the absolute ambipolar mobility (m/s-N) and $B_{amb} = D_{amb}/RT$, D_{amb} = ambipolar diffusion coefficient (m^2/s) and R and T have their usual meanings. Of course, the fluxes in Eqn. (4) are all equal because of stoichiometry. The goal is to determine D_{amb} in terms of the tracer diffusion coefficients of magnesium and oxygen, D^*_{Mg} and D^*_O. In general, both J and $\nabla\mu_{MgO}$ are vectors but that is ignored, again for simplicity.

Chemical Potentials and Fluxes

For MgO,

$$Mg(g) + 1/2\ O_2(g) = MgO(s) \tag{5}$$

$$\mu_{Mg(g)} + \tfrac{1}{2}\mu_{O_2(g)} = \mu_{MgO(s)} \tag{6}$$

and these chemical potentials, μ_i (J/mole), are all measurable and definable. However, the solid-state diffusion of magnesium and oxygen in MgO are effected by the coupling of their respective ions to maintain charge neutrality (assuming negligible electronic conductivity). Now, the relations between measurable and lattice species is given by

$$\begin{aligned} Mg_{Mg}^{X} &= Mg(g) + V_{Mg}'' + 2h^{\bullet} \\ O_{O}^{X} &= \tfrac{1}{2}O_2(g) + V_{O}^{\bullet\bullet} + 2e' \end{aligned} \tag{7}$$

with the corresponding equations for the chemical potentials

$$\begin{aligned} \mu_{Mg^{+2}} &= \mu_{Mg} + \mu_{V_{Mg}''} + 2\mu_{h^{\bullet}} \\ \mu_{O^{-2}} &= \tfrac{1}{2}\mu_{O_2} + \mu_{V_O^{\bullet\bullet}} + 2\mu_{e'} \end{aligned} \tag{8}$$

In an electric field, E, the movement of both Mg^{+2} and O^{-2} ions are affected by the field:

$$\begin{aligned} J_{MgO} &= J_{Mg} = J_{Mg^{+2}} = -C_{Mg}B_{Mg}\left(\nabla\mu_{Mg^{+2}} + 2eE\right) \\ &\text{and} \\ J_{MgO} &= J_{O} = J_{O^{-2}} = -C_{O}B_{O}\left(\nabla\mu_{O^{-2}} - 2eE\right) \end{aligned} \tag{9}$$

If there is no external field, then an internal field is generated that couples the two diffusing species and is obtained by equating the two equations in Eqn. (9) since the fluxes of magnesium and oxygen must be equal to preserve neutrality, Eqn. (4), giving:

$$2eE = \frac{D_O^{*}\nabla\mu_{O^{-2}} - D_{Mg}^{*}\nabla\mu_{Mg^{+2}}}{D_O^{*} + D_{Mg}^{*}} \tag{10}$$

since $C_O = C_{Mg} = C_{MgO}$ and $B_i = D_i^{*}/RT$. Adding the two equations in Eqn. (8) and taking the gradient gives:

$$\nabla\mu_{Mg^{+2}} + \nabla\mu_{O^{-2}} = \nabla\left(\mu_{Mg} + \tfrac{1}{2}\nabla\mu_{O_2}\right) = \nabla\mu_{MgO} \tag{11}$$

since $\mu_{V_{Mg}''} + \mu_{V_O^{\bullet\bullet}} = 0$ by definition for vacancies in thermal equilibrium,[1] and $\mu_{e'} = E_F$ and $\mu_{h^{\bullet}} = -E_F$ where E_F is the Fermi energy for electrons and holes. Combination of the first of Eqn. (9) with Eqns. (10) and (11) gives the final result:

$$J_{MgO} = -\frac{C_{MgO}}{RT}\left(\frac{D_{Mg}^{*}D_O^{*}}{D_{Mg}^{*} + D_O^{*}}\right)\nabla\mu_{MgO} \tag{12}$$

resulting in

$$D_{amb} = \left(\frac{D_{Mg}^{*}D_O^{*}}{D_{Mg}^{*} + D_O^{*}}\right) \tag{13}$$

for MgO and in the general case of a compound $A_\alpha B_\beta$:[1]

$$D_{amb} = \left(\frac{D_A^* D_B^*}{\beta D_A^* + \alpha D_B^*} \right). \tag{14}$$

NECK GROWTH BY VAPOR TRANSPORT

Thermodynamics

Consider neck growth between particles of MgO during sintering by vapor transport through the gas phase, which of course only leads to neck growth and no shrinkage.[8,9] Again, for the sake of concreteness, assume that the vapor transport is enhanced by an HCl gas atmosphere:[10]

$$MgO(s) + 2\ HCl(g) = MgCl_2(g) + H_2O(g). \tag{15}$$

The free energy of the solid MgO is altered by the neck curvature, κ;

$$\Delta G(MgO) = \Delta G^{o}(MgO) - \kappa\gamma\Omega, \tag{16}$$

so that the free energy for the above reaction is given by:

$$\Delta G_R = \Delta G^{o}(MgCl_2) + \Delta G^{o}(H_2O) - 2\Delta G^{o}(HCl) - \Delta G(MgO) \tag{17}$$

which becomes

$$\Delta G_R = \Delta G_R^{o} + \kappa\gamma\Omega \tag{18}$$

where ΔG_R^{o} is the free energy for the reaction without curvature. This leads to the equilibrium constant:

$$\frac{p(MgCl_2)p(H_2O)}{p(HCl)^2} = K_p e^{\frac{\kappa\gamma\Omega}{RT}} \tag{19}$$

where K_p is the equilibrium constant without curvature and, since $\kappa\gamma\Omega/RT$ is usually small,

$$p(MgCl_2)p(H_2O) \simeq p(HCl)^2 K_p \left(1 + \frac{\kappa\gamma\Omega}{RT}\right). \tag{20}$$

To simplify what follows, let

$$p(MgCl_2) = p_M, \quad p(H_2O) = p_H, \quad p(HCl)^2 K_p = A, \quad \text{and} \ \frac{\kappa\gamma\Omega}{RT} = \beta \tag{21}$$

so $p_M p_H = A(1+\beta)$ and $p_M^e = p_H^e = A^{\frac{1}{2}}$, which are the equilibrium pressures of $MgCl_2$ and H_2O in the absence of curvature assuming that all of the water vapor in the system comes from the reaction; i.e. the ambient water vapor is very low.

Coupled Diffusion for MgO

In the cases of neck growth by vapor phase transport, it is the deviation from the equilibrium pressures caused by curvature that leads to mass transport. For neck growth in MgO, the fluxes of $MgCl_2$, J_M, and H_2O, J_H, can be given by:

$$J_M = -\frac{gD_M}{RT}\left(p_M - p_M^e\right)$$
$$J_H = -\frac{gD_H}{RT}\left(p_H - p_H^e\right) \tag{22}$$

where D_M and D_H are the gas diffusion coefficients for $MgCl_2$ and H_2O and the equilibrium pressures $p_M^e = p_H^e = A^{\frac{1}{2}}$ far from the neck where $\kappa \simeq 0$ and g = some geometric factor as above [the RT in the denominator comes from expressing gas concentrations (mole/m^3) in terms of pressures (MPa)]. Again, the two fluxes must be equal, $J_{MgO} = J_M = J_H$, because MgO must be transported; i.e. stoichiometry must be maintained. This coupling along with Eqn.(20) leads to a quadratic equation for p_M:

$$\frac{D_M}{D_H}p_M^2 + A^{\frac{1}{2}}\frac{(D_H - D_M)}{D_H}p_M - A(1+\beta) = 0 \tag{23}$$

and since β is small, Eqn. (23) can be solved to give:

$$p_M \simeq A^{\frac{1}{2}}\left(1 + \frac{D_H}{D_M + D_H}\beta\right) \tag{24}$$

and from Eqns.(22) and (24)

$$p_H \simeq A^{\frac{1}{2}}\left(1 + \frac{D_M}{D_M + D_H}\beta\right) \tag{25}$$

so that the product of Eqns.(24) and (25) is Eqn. (20), neglecting terms in β^2 since β is small. Eqns. (24) and (25) show the relative amount that each pressure is raised over the normal equilibrium vapor pressure depends on the ratio of the diffusion coefficients. Substitution of Eqn. (24) [or Eqn. (25)] into Eqn. (22) gives the fluxes of MgO, $MgCl_2$, and H_2O:

$$J_{MgO} = J_{MgCl_2} = J_{H_2O} = -\frac{gA^{\frac{1}{2}}}{RT}\frac{\beta}{2}\left(\frac{2D_{MgCl_2}D_{H_2O}}{D_{MgCl_2} + D_{H_2O}}\right) = -\frac{gp^e_{MgCl_2}}{RT}\frac{\kappa\gamma\Omega}{2RT}\left(\frac{2D_{MgCl_2}D_{H_2O}}{D_{MgCl_2} + D_{H_2O}}\right). \tag{26}$$

with the effective coupled diffusion coefficient now given by:

$$D_{eff} = \frac{2D_{MgCl_2}D_{H_2O}}{D_{MgCl_2} + D_{H_2O}}. \tag{27}$$

It should be noted that if $D_{MgCl_2} = D_{H_2O}$ then $D_{eff} = D_{MgCl_2} = D_{H_2O}$ and, from Eqns. (24) and (25) $p_{MgCl_2} = p_{H_2O} = A^{\frac{1}{2}}\left(1+\frac{\beta}{2}\right)$ as is required by Eqn. (20).

General Case for a Compound $A_\alpha O_\beta$ in HCl Gas

For the general case of an oxide compound $A_\alpha O_\beta$ being transported by HCl gas:

$$A_\alpha O_\beta(s) + 2\beta\, HCl(g) = \alpha\, ACl_\beta(g) + \beta\, H_2O(g) \tag{28}$$

the coupled flux equations become:[11]

$$J_{A_\alpha B_\beta} = -\frac{1}{\alpha}\frac{gp^e_{ACl_\beta}}{RT}\frac{\kappa\gamma}{(\alpha+\beta)}\frac{\Omega}{RT}\left(\frac{2D_{ACl_\beta}D_{H_2O}}{D_{ACl_\beta}+D_{H_2O}}\right) = \frac{J_{ACl_\beta}}{\alpha}$$

$$J_{A_\alpha B_\beta} = -\frac{1}{\beta}\frac{gp^e_{H_2O}}{RT}\frac{\kappa\gamma}{(\alpha+\beta)}\frac{\Omega}{RT}\left(\frac{2D_{ACl_\beta}D_{H_2O}}{D_{ACl_\beta}+D_{H_2O}}\right) = \frac{J_{H_2O}}{\beta} \tag{29}$$

and the effective gas diffusion coefficient has the same form as it does for MgO: it is twice the harmonic mean of the two gas diffusion coefficients.

ACTIVE GAS CORROSION

Corrosion of MgO in HCl Gas

In this case, consider a sphere of MgO undergoing active gas corrosion in pure HCl with a very low ambient water vapor pressure, Figure 2 and Eqn. (15). This implies that all of the water vapor in the atmosphere comes from the reaction so that again $p^e_{MgCl_2} = p^e_{H_2O}$. Here the effect of surface curvature is neglected and the equilibrium constant, Eqn. (15).is:

$$\frac{p(MgCl_2)p(H_2O)}{p(HCl)^2} = K_p \tag{30}$$

For a sphere, the concentration as a function of the sphere radius, a, and the radial distance, r, is given by $C = C_0\, a/r$ as shown in Figure 2. Therefore, fluxes of $MgCl_2$, H_2O, and MgO at the surface of the sphere, r = a, are given by (since in the gas, C = p/RT):

$$J_{MgO} = J_{MgCl_2} = \frac{D_{MgCl_2}}{aRT}p^0_{MgCl_2} = J_{H_2O} = \frac{D_{H_2O}}{aRT}p^0_{H_2O} \tag{31}$$

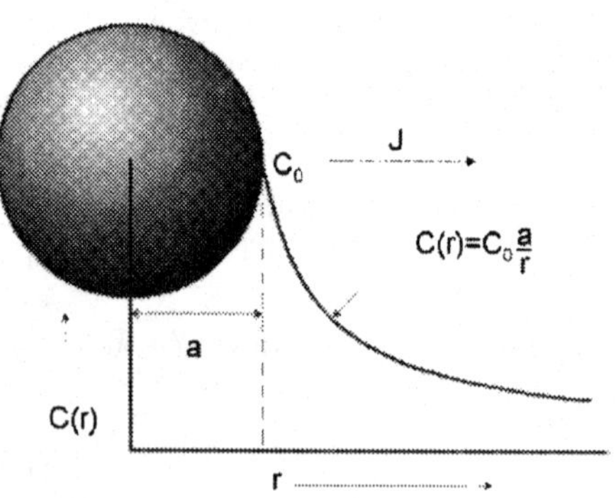

Figure 2. Sphere undergoing active gas corrosion

Substitution of Eqn. (31) into Eqn. (30) gives:

$$
\begin{aligned}
p^{0}_{MgCl_2} &= \left(p^{2}_{HCl}K_p\right)^{\frac{1}{2}}\left(\frac{D_{H_2O}}{D_{MgCl_2}}\right)^{\frac{1}{2}} = p^{e}_{MgCl_2}\left(\frac{D_{H_2O}}{D_{MgCl_2}}\right)^{\frac{1}{2}} \\
p^{0}_{H_2O} &= \left(p^{2}_{HCl}K_p\right)^{\frac{1}{2}}\left(\frac{D_{H_2O}}{D_{MgCl_2}}\right)^{\frac{1}{2}} = p^{e}_{H_2O}\left(\frac{D_{MgCl_2}}{D_{H_2O}}\right)^{\frac{1}{2}}
\end{aligned}
\tag{32}
$$

and Eqns. (32) back into Eqns. (31) gives:

$$
\begin{aligned}
J_{MgO} = J_{MgCl_2} &= -\frac{1}{aRT}\left(D_{MgCl_2}D_{H_2O}\right)^{\frac{1}{2}} p^{e}_{MgCl_2} \\
J_{MgO} = J_{H_2O} &= -\frac{1}{aRT}\left(D_{MgCl_2}D_{H_2O}\right)^{\frac{1}{2}} p^{e}_{H_2O}
\end{aligned}
\tag{33}
$$

Eqn. (32) shows that the surface pressures are not the same but are adjusted to take into consideration the differences in the gas diffusion coefficients of $MgCl_2$ and H_2O to ensure that there is a stoichiometric flux of MgO away from the sphere. Eqn. (33) shows the effective diffusion coefficient is the geometric mean of the diffusion coefficients of the two gases.

General Case for a Compound $A_\alpha O_\beta$ in HCl Gas

For the general case of an oxide compound $A_\alpha O_\beta$ being transported by HCl gas:

$$
A_\alpha O_\beta(s) + 2\beta\, HCl(g) = \alpha\, ACl_\beta(g) + \beta\, H_2O(g) \tag{34}
$$

the coupled flux equations become:[11]

$$
\begin{aligned}
J_{A_\alpha B_\beta} &= \frac{J_{ACl_\beta}}{\alpha} = -\frac{1}{\alpha aRT}\left(D^{\alpha}_{ACl_\beta}D^{\beta}_{H_2O}\right)^{\frac{1}{\alpha+\beta}} p^{e}_{ACl_\beta} \\
J_{A_\alpha B_\beta} &= \frac{J_{H_2O}}{\beta} = -\frac{1}{\beta aRT}\left(D^{\alpha}_{ACl_\beta}D^{\beta}_{H_2O}\right)^{\frac{1}{\alpha+\beta}} p^{e}_{H_2O}
\end{aligned}
\tag{35}
$$

For active gas corrosion, the apparent diffusion coefficient is a mole-weighted geometric mean of the diffusion coefficients of the transporting gases.

CONCLUSIONS

Mass transport of a compound to maintain either electrical neutrality or stoichiometry leads to coupled diffusion coefficients that are some kind of mean of the constituent diffusion coefficients. The type of mean depends on the transport process.

REFERENCES

[1] D. W. Readey, "Chemical Potentials and Initial Sintering in Pure Metals and Ionic Compounds," *Journal of Applied Physics,* **37** [6] 2309-2312 (1966).

[2] D. W. Readey, "Mass Transport and Sintering in Impure Ionic Solids," *Journal of the American Ceramic Society,* **49** [7] 366-369 (1966).

[3] F. A. Kröger, *The Chemistry of Imperfect Crystals.* North-Holland Publishing Co., Amsterdam, 1964.

[4] J. S. Blakemore, *Semiconductor Statistics.* Pergamon Press, New York, 1962.

[5] Y-M. Chiang, D. B. Birnie, III, and W. D. Kingery, *Physical Ceramics.* Wiley, New York, 1997.

[6] M. N. Rahaman, *Ceramic Processing and Sintering.* Marcel Dekker, New York, 1995.

[7] J. W. Evans and L. C. DeJonghe, *The Production of Inorganic Materials.* Macmillan, New York, 1991.

[8] G. C. Kuczynski, "Self-Diffusion in Sintering of Metallic Particles," *Transactions of AIME,* **185** [2] 169-178 (1949).

[9] W. D. Kingery and M. Berg, "Study of Initial Stage of Sintering Solids by Viscous Flow, Evaporation-Condensation, and Self-Diffusion," *Journal of Applied Physics.* **26** [10] 1205-1212 (1955).

[10] D. W. Readey, "Vapor Transport and Sintering," pp. 86-110 in Sintering of Advanced Ceramics, Ceramic Transactions, Vol. 7, C. A. Handwerker, J. E. Blendell, and W. A. Kaysser, eds., (The Am. Ceramic Society, Columbus), 1990.

[11] D. W. Readey and C. M. Tuell, "Modeling of Diffusion Fluxes During Grain Boundary Grooving and Active Gas Corrosion," submitted for publication.

Author Index

Keyword Index

9 781574 981780